手写Spring
渐进式源码实践

付政委（小傅哥）著

电子工业出版社
Publishing House of Electronics Industry
北京·BEIJING

内 容 简 介

本书基于 Spring 框架的核心逻辑，通过实现简化版 Spring 框架的方式，对 Spring 进行分析、设计和实践。本书以实践为核心，摒弃 Spring 源码中繁杂的内容，选择框架中的核心逻辑，简化代码实现过程，保留核心功能，如 IOC、AOP、Bean 的生命周期、上下文、作用域、资源处理、JDBC、事务、开发简易版 ORM 框架、将 ORM 框架整合到 Spring 框架中等内容的方案设计和源码实现。在 Spring 框架功能模块的开发过程中，逐步介绍并深入分析其中所涉及的设计原则和设计模式，使读者游刃有余地解决在调试 Spring 源码及开发 SpringBoot Starter 组件时遇到的问题。

本书既可以作为计算机相关行业研发人员的辅导书，也可以作为高等院校计算机专业学生的参考书。无论是初学者，还是中、高级研发人员，都能从本书中获得启发。

未经许可，不得以任何方式复制或抄袭本书之部分或全部内容。
版权所有，侵权必究。

图书在版编目（CIP）数据

手写 Spring：渐进式源码实践 / 付政委著. —北京：电子工业出版社，2022.11
ISBN 978-7-121-44420-3

Ⅰ. ①手… Ⅱ. ①付… Ⅲ. ①JAVA 语言－程序设计 Ⅳ. ①TP312.8

中国版本图书馆 CIP 数据核字（2022）第 191545 号

责任编辑：宋亚东　　　　　　　　　特约编辑：田学清
印　　　刷：北京东方宝隆印刷有限公司
装　　　订：北京东方宝隆印刷有限公司
出版发行：电子工业出版社
　　　　　北京市海淀区万寿路 173 信箱　　邮编：100036
开　　本：787×980　　1/16　　印张：19.5　　字数：391 千字
版　　次：2022 年 11 月第 1 版
印　　次：2022 年 11 月第 2 次印刷
定　　价：128.00 元

凡所购买电子工业出版社图书有缺损问题，请向购买书店调换。若书店售缺，请与本社发行部联系，联系及邮购电话：（010）88254888，88258888。
质量投诉请发邮件至 zlts@phei.com.cn，盗版侵权举报请发邮件至 dbqq@phei.com.cn。
本书咨询联系方式：（010）51260888-819，faq@phei.com.cn。

前　言

　　Spring 是一个优秀的轻量级 Java 开发框架。2003 年 6 月，Rod Johnson 创建了 Spring 框架，第一次发布在 Apache 2.0 许可证下，后将其改名为 Spring 并且正式发布。

　　2002 年 10 月，Rod Johnson 在 *Expert One-on-One J2EE Design and Development* 一书中，对 Java EE 系统的框架臃肿、低效、脱离现实等问题提出了质疑，并提出了一个基于普通 Java 类和依赖注入的简单解决方案。然后，Rod Johnson 创建了 Interface 21 框架。它是一个从实际需求出发，着重于轻便、灵巧，易于开发、测试和部署的轻量级开发框架。Spring 框架以 Interface 21 框架为基础，经过重新设计和内容的不断丰富，里程碑式的 Spring 1.0 版本于 2004 年 3 月正式发布。同年，Rod Johnson 又推出了一部堪称经典的图书 *Expert one-on-one J2EE Development without EJB*，该书在 Java 开发领域掀起了轩然大波，不断改变着 Java 开发者对程序设计和开发的思考方式。

　　对于希望长期从事编程开发的架构师和技术研发人员，不仅要熟练使用 Spring 框架，还要掌握其源码设计，通过学习这些优秀的设计思想和设计模式在实际场景中的应用方法，提高自己的业务工程架构设计能力，以及对基于 Spring 框架所实现的一些技术类组件的驾驭与把控能力。

为什么撰写本书

　　大部分使用 Spring 框架的研发人员在遇到 Spring 框架的报错提醒，以及需要基于 Spring 框架开发 SpringBoot Starter 等技术类组件时，会尝试阅读 Spring 框架的源码。由于 Spring 框架的源码体量庞大、语法复杂，也不像平常的业务流程开发代码一样具有分层结构，并且其中使用了大量的设计模式，所以阅读难度较大。研发人员很难厘清其中的调用链路和各个类之间的关系。

笔者阅读了不少关于 Spring 的图书，在反复学习源码后，仍然难以理解 Spring 框架中各个功能的实现细节。其中一个原因是自己没有动手实践，只阅读了图书，很难完全掌握 Spring 框架的精髓。因此，笔者结合《Spring 源码深度解析》《精通 Spring 4.x——企业应用开发实战》等图书，以及 GitHub 中的优秀案例——code4craft/tiny-spring、DerekYRC/mini-spring 等，逐步完成了 Spring 框架中核心源码的实现，并扩展了 ORM 框架。在编写的过程中，还收到了 GitHub 用户 zh-d-d、cwjokaka、numgin 提交的 PR，在此表示感谢。在学习过程中，笔者对 Spring 框架有了非常深入的了解和认识，也体会了更多精妙的设计原则和设计模式。所以，我把关于手动实现简单版 Spring 框架的内容编写成书，希望可以帮助更多的研发人员学习 Spring 源码，编写出有价值的设计方案。

本书主要内容

本书采用从零手写 Spring 的方式，摒弃 Spring 源码中繁杂的内容，选择整体框架中的核心逻辑，简化代码实现过程，保留核心功能，如 IOC、AOP、Bean 的生命周期、上下文、作用域和资源处理等。在开发过程中，细化功能模块，逐步完成一个简单版的 Spring 框架。

本书共 21 章。引言介绍了学习 Spring 的方法，以及本书源码的获取和使用方式。

- 第 1～10 章：主要介绍 IOC 容器，逐步完善一个简单的 Spring Bean 容器的相关功能，引入实例化策略、注入属性和依赖、设计应用上下文、处理 Bean 对象的生命周期，以及实现感知容器对象的监听等。

- 第 11～12 章：主要介绍 AOP 切面，基于 JDK、Cglib 的动态代理、方法拦截、切点表达式等技术，将代理与 Spring Bean 容器整合，提供 AOP 切面功能。

- 第 13～17 章：扩展简单版 Spring 框架的自动化功能，完成自动扫描注册、注解和代理注入，以及通过三级缓存处理对象的循环依赖等功能。

- 第 18～21 章：基于简单实现的 Spring 框架整合 JDBC、事务的功能，开发一个简单版的 ORM 框架，并将 ORM 框架整合到 Spring Bean 容器中，介绍自定义代理对象的扫描和注册过程。

如何阅读本书

本书主要通过渐进式开发功能模块，以实现整个 Spring 框架的核心源码开发。首先，每章开头都会列出难度和重点；然后，正文中会介绍要处理的问题、具体设计和实现代码；最后，给出测试验证和本章总结。

在阅读本书的过程中，建议读者先阅读"引言"，以便从全局的视角了解本书要实现的 Spring 框架的内容，掌握学习方法。同时，"引言"中也列举了本书工程源码的环境配置、获取和使用方法。

致谢

要特别感谢我的父母（付井海、徐文杰）、妻子（郭维清），由于他们在日常生活中分担了许多家庭任务，才让我有更多的时间投入文字创作中，使得这本书能与广大读者见面。

感谢灵魂设计师 Beebee 老师为本书设计封面插图。

博文视点的宋亚东编辑的热情推动最终促成了我与电子工业出版社的合作。感谢电子工业出版社博文视点对本书的重视，以及为本书出版所做的一切。

由于笔者水平有限，书中难免存在一些疏漏和不足，希望广大同行专家和读者给予批评与指正。

付政委（小傅哥）

读者服务

微信扫码回复：44420

- 获取本书配套源码资源。
- 加入本书读者交流群，与更多读者互动。
- 获取【百场业界大咖直播合集】（持续更新），仅需 1 元。

目 录

引言 / 1

第 1 章 实现一个简单的 Spring Bean 容器 / 7

1.1 容器是什么 / 7
1.2 简单容器设计 / 8
1.3 简单容器实现 / 9
1.4 容器使用测试 / 12
1.5 本章总结 / 13

第 2 章 实现 Bean 对象的定义、注册和获取 / 14

2.1 容器的思考 / 14
2.2 完善容器设计 / 15
2.3 完善容器实现 / 16
2.4 Bean 生命周期测试 / 22
2.5 本章总结 / 24

第 3 章 基于 Cglib 实现含构造函数的类实例化策略 / 25

3.1 实例化问题 / 25
3.2 实例化策略设计 / 26
3.3 实例化策略代码实现 / 27
3.4 构造函数对象测试 / 32
3.5 本章总结 / 35

第 4 章 注入属性和依赖对象 / 36

4.1 Bean 对象拆解思考 / 36
4.2 属性填充设计 / 37
4.3 属性填充实现 / 38
4.4 注入属性测试 / 43
4.5 本章总结 / 46

第 5 章 资源加载器解析文件注册对象 / 47

5.1 对象创建问题 / 47
5.2 资源加载和解析设计 / 48
5.3 资源加载和解析设计实现 / 49
5.4 配置 Bean 对象注册测试 / 60
5.5 本章总结 / 64

第 6 章 实现应用上下文 / 65

- 6.1 分治 Bean 对象功能 / 65
- 6.2 Bean 对象扩展和上下文设计 / 66
- 6.3 Bean 对象扩展和上下文实现 / 68
- 6.4 应用上下文功能测试 / 79
- 6.5 本章总结 / 83

第 7 章 Bean 对象的初始化和销毁 / 84

- 7.1 容器管理 Bean 功能 / 84
- 7.2 初始化和销毁设计 / 85
- 7.3 初始化和销毁实现 / 87
- 7.4 容器功能测试 / 97
- 7.5 本章总结 / 99

第 8 章 感知容器对象 / 101

- 8.1 Spring Bean 容器的功能 / 101
- 8.2 感知容器设计 / 102
- 8.3 感知容器实现 / 103
- 8.4 Aware 接口的功能测试 / 111
- 8.5 本章总结 / 114

第 9 章 对象作用域和 FactoryBean / 117

- 9.1 Bean 对象的来源和模式 / 117
- 9.2 FactoryBean 和对象模式设计 / 118
- 9.3 FactoryBean 和对象模式实现 / 119
- 9.4 代理 Bean 和对象模式测试 / 128
- 9.5 本章总结 / 132

第 10 章 容器事件和事件监听器 / 134

- 10.1 运用事件机制降低耦合度 / 134
- 10.2 事件观察者设计 / 135
- 10.3 事件观察者实现 / 136
- 10.4 事件使用测试 / 146
- 10.5 本章总结 / 148

第 11 章 基于 JDK、Cglib 实现 AOP 切面 / 150

- 11.1 动态代理 / 150
- 11.2 AOP 切面设计 / 151
- 11.3 AOP 切面实现 / 152
- 11.4 AOP 切面测试 / 163
- 11.5 本章总结 / 165

第 12 章 把 AOP 融入 Bean 的生命周期 / 166

- 12.1 AOP 与框架整合思考 / 166

12.2　AOP 切面设计 / 167
12.3　AOP 切面实现 / 168
12.4　切面使用测试 / 176
12.5　本章总结 / 179

第 13 章　自动扫描注册 Bean 对象 / 180

13.1　注入对象完善点 / 180
13.2　自动扫描注册设计 / 181
13.3　自动扫描注册实现 / 182
13.4　注册 Bean 对象测试 / 190
13.5　本章总结 / 193

第 14 章　通过注解注入属性信息 / 194

14.1　引入注入注解 / 194
14.2　注入属性信息设计 / 195
14.3　注入属性信息实现 / 196
14.4　注解使用测试 / 206
14.5　本章总结 / 208

第 15 章　给代理对象设置属性注入 / 210

15.1　代理对象创建过程问题 / 210
15.2　代理对象属性填充设计 / 211
15.3　代理对象属性填充实现 / 212
15.4　代理对象属性注入测试 / 219
15.5　本章总结 / 221

第 16 章　通过三级缓存解决循环依赖 / 222

16.1　复杂对象的创建思考 / 222
16.2　循环依赖设计 / 223
16.3　循环依赖实现 / 227
16.4　循环依赖测试 / 234
16.5　本章总结 / 238

第 17 章　数据类型转换 / 239

17.1　类型转换设计 / 239
17.2　类型转换实现 / 240
17.3　类型转换测试 / 249
17.4　本章总结 / 251

第 18 章　JDBC 功能整合 / 252

18.1　JdbcTemplate 说明 / 252
18.2　整合 JDBC 服务设计 / 253
18.2　整合 JDBC 服务开发 / 254
18.3　JDBC 功能测试 / 258
18.4　本章总结 / 260

第 19 章　事务处理 / 261

19.1　了解事务 / 261
19.2　事务功能设计 / 262
19.3　事务功能实现 / 264

19.4 切面事务测试 /272
19.5 本章总结 /276

第 20 章 ORM 框架实现 /278

20.1 简单 ORM 框架设计 /278
20.2 简单 ORM 框架实现 /279
20.3 ORM 框架使用测试 /287
20.4 本章总结 /290

第 21 章 将 ORM 框架整合到 Spring Bean 容器中 /291

21.1 ORM-Spring 整合设计 /291
21.2 ORM-Spring 整合实现 /292
21.3 整合功能验证 /299
21.4 本章总结 /302

引言

为什么学 Spring

Java 程序员几乎离不开通过 Spring 构建起来的开发技术栈，其中，需要用到的框架包括 Spring、SpringBoot、SpringMVC 和 MyBatis 等。从事程序开发的时间越久，就越需要更加深入地了解这些框架的内核，以便逐步从初级码农晋升到高级研发人员甚至架构师时，能凭自身的技术能力，解决一些复杂的框架应用问题，如结合 SpringBoot 框架的 Starter 技术进行组件设计和开发、依赖 SPI 机制扩展业务服务、处理一些棘手问题等。

如果想熟练地解决复杂的设计和应用问题，而不只是做一些 API 的应用和开发工作，就需要对 Spring 的框架源码有更深入的了解。作为研发人员，不应该只是通过机械记忆的方式来学习一些碎片化的知识点，如对象的作用域、注册方式、销毁方式、Bean 工厂、通过三级缓存解决循环依赖等。因为这些碎片化的知识点并不能完整地串联出整个 Spring 框架的核心脉络，随着时间的流逝，机械记忆的知识点也会被慢慢遗忘。

怎么学好 Spring

学习 Spring 的最佳方式应该是按照 Spring 框架的知识体系，拆分出核心流程并进行开发实践，就像开发一个项目一样，渐进式地完善各个模块的功能。我们通过实践既可以非常熟练地掌握 Spring 框架的技术，也可以学习怎样更好地运用设计模式，并能在阅读 Spring 源码时，看清核心流程之外的扩展，把能力运用到实际的项目开发中。

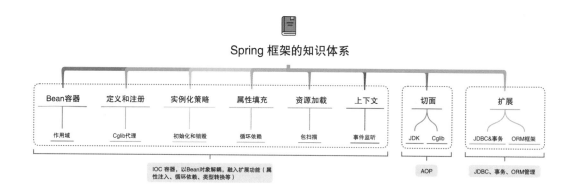

本书以实践为主，避免重复讲解 Spring 源码或只讲解理论。读者通过摒弃 Spring 源码中的繁杂内容，抽取核心逻辑，简化代码，以渐进式的开发方式，逐步实现 Spring 框架中的核心功能，既包括以一横一纵 IOC、AOP 的框架结构完成 Bean 对象的完整生命周期、上下文、作用域、资源处理和循环依赖等内容，也包括扩展整合 JDBC、事务、ORM 框架等内容，可以逐步看到 Spring 框架从无到有的形成过程。

在开发实践项目的过程中，如果能跟随章节内容学习到最后，则阅读 Spring 源码时会更加容易，也能更好地掌握 Spring 框架的知识体系。

Spring 框架地图

本书以 Bean 的生命周期为核心，逐步完成各个功能节点的开发，以及学习如何运用 Spring 框架拓展链接外部服务。

读者在学习过程中，可以参考 Spring 框架地图通过全局的视角，可以更好地理解和学习 Spring 框架的设计与开发。

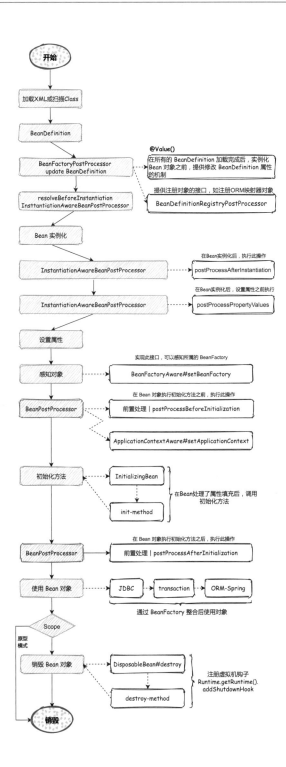

工程使用说明

此工程以 IntelliJ IDEA + Maven 的方式搭建，共计 21 个开发工程模块，逐步实现 Spring 框架的核心功能。在学习过程中，读者可以根据章节提示，打开对应的项目工程进行开发和调试。

读者需要根据不同平台的源码地址选择一个可以访问的路径下载源码。

1. 源码地址

工程源码地址分别由 GitHub、GitCode 两个平台提供，而读者可以选择任意一个平台下载工程源码。

- GitHub：https://github.com/fuzhengwei/book-small-spring。
- Gitcode：https://gitcode.net/fuzhengwei/book-small-spring。

2. 环境配置

对于环境配置，各版本的情况为：JDK 1.8.x，不建议使用过高版本的 JDK；Maven 3.6.x；MySQL 5.x 或 MySQL 8.x 都可以。在调试时注意配置对应的驱动程序 com.mysql.jdbc.Driver 或 com.mysql.cj.jdbc.Driver，以及 IntelliJ IDEA Community Edition 2020.x。环境所需的其他版本都已经在对应工程的 POM 文件中设置完成，如果开发过程中需要的某个版本在 Maven 仓库中不存在，则需要替换。

3. 工程结构

Spring 框架中共包含 21 个开发工程模块。读者在学习过程中，可以在根目录中使用 IDEA 工具打开已经下载的代码，或者在每次学习新章节时，打开本章对应的源码。

> 注意：当首次打开工程，未把项目加载到 Maven 工程时，可以在工程的 POM 文件中单击"Add as Maven Project"按钮，完成初始化。

```
SpringTutorials ~/1024/SpringTutorials
  .idea
  docs
  spring-step-01
  spring-step-02
  spring-step-03
  spring-step-04
  spring-step-05
  spring-step-06
  spring-step-07
  spring-step-08
  spring-step-09
  spring-step-10
  spring-step-11
  spring-step-12
  spring-step-13
  spring-step-14
  spring-step-15
  spring-step-16
  spring-step-17
  spring-step-18
  spring-step-19
  spring-step-20
  spring-step-21
```

更好的开始

1. 前置条件

读者需要具备一定的编码基础，对 IDEA、JDK、Maven 等工具的配置方法较为熟悉，使用 Spring 框架开发过项目，并能够排查一些简单的源码问题。如果读者尚未掌握以上内容，则建议补充相关知识，否则阅读时会有些吃力。

2. 能学到什么

- 看得懂：Bean 容器是如何定义和实现的。
- 学得会：工厂模式、策略模式、观察者模式等是如何在 Spring 框架中体现的。

- 厘得清：从应用上下文到创建 Bean 对象，串联出完整的生命周期。
- 吃得透：IOC、AOP、代理、切面、循环依赖等都是如何设计和实现的。
- 玩得转：自研框架时，如何自动注册对象并交给 Spring Bean 容器管理。

3. 学习建议

本书基于开发简化版 Spring 框架，介绍 Spring 框架的原理和核心知识，不仅注重代码的编写与实现，而且注重内容的需求分析和方案设计。读者在学习过程中要结合这些内容来实践并调试对应的代码，在学习过程中遇到问题是正常的，在坚持亲自解决问题后，一定会对各个知识点有更深刻的体会和认识。

第 1 章
实现一个简单的 Spring Bean 容器

面对复杂的源码，试着找到开头和结尾是一件非常具有挑战的事。为了让更多的初学者上手，从本章开始，我们将通过实践的方式带领读者逐步实现 Spring 框架的核心链路和功能逻辑。

简化上手、凸显重点、摒弃冗余，让更多的初学者都能在这场学习旅途中收获满满。旅途即将开始，你准备好了吗？

- 本章难度：★☆☆☆☆
- 本章重点：基于 Spring Bean 容器的存储功能和读取功能，采用时间复杂度为 $O(\log n)$ 的 HashMap 数据结构进行设计和实现。

1.1 容器是什么

Spring Bean 包含并管理应用对象的配置和生命周期。从这个意义上讲，它是一种用于承载对象的容器，开发者可以设置每个 Bean 对象是如何被创建的，以及它们是如何互相关联、构建和使用的。使用这些 Bean 对象可以创建一个单独的实例，或者在需要时生成一个新的实例。

如果将一个 Bean 对象交给 Spring Bean 容器管理，则这个 Bean 对象会以类似零件的方式被拆解，然后存储到 Spring Bean 容器的定义中，便于 Spring Bean 容器管理，这相当于把对象解耦。

当一个 Bean 对象被定义和存储后，会由 Spring Bean 容器统一进行分配，这个过程包括 Bean 对象的初始化和属性填充等。最终，我们可以完整地使用一个被实例化后的 Bean 对象。

本章将带领读者实现一个简单的 Spring Bean 容器，用于定义、存储和获取 Bean 对象。

1.2 简单容器设计

我们将可以存储数据的数据结构称为容器，如 ArrayList、LinkedList、HashSet 等。但在 Spring Bean 容器中，需要一种可以用于存储对象和使用对象名称进行便捷索引的数据结构，所以选择 HashMap 数据结构是最合适的，下面简单介绍 HashMap。

HashMap 是一种基于扰动函数、负载因子和红黑树转换等技术形成的拉链寻址的数据结构，能让数据更加均匀地分布在哈希桶，以及碰撞时形成的链表和红黑树上。HashMap 的数据结构会最大限度地使整个数据读取的复杂度范围为 $O(1) \sim O(n)$，也存在较多使用复杂度为 $O(n)$ 的链表查找数据的情况。不过，经过 10 万个单词数据的扰动函数索引计算后，通过在寻址位置膨胀的方差稳定性对比验证得出，使用扰动函数的数据会更均匀地分布在各个哈希桶索引上，基于这些特性的 HashMap 非常适合用于实现 Spring Bean 容器。

另外，实现一个简单的 Spring Bean 容器，还需要完成 Bean 对象的定义、注册和获取 3 个基本步骤，如图 1-1 所示。

1. 定义

BeanDefinition 是我们在查阅 Spring 源码时经常看到的一个类，如 singleton 属性、prototype 属性和 BeanClassName 类型等。这里的实现会采用更加简单的方法，只需要定义一个 Object 类型用于存储任意类型的对象。

2. 注册

注册过程相当于把数据存储到 HashMap 中，现在 HashMap 中存储的是被定义的 Bean 对象的信息。

第 1 章　实现一个简单的 Spring Bean 容器

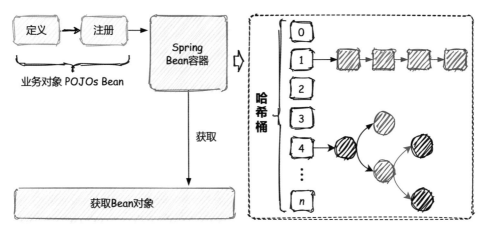

图 1-1

3. 获取

最后就是获取对象。Bean 对象的名字就是 key。当 Spring Bean 容器初始化 Bean 对象后，Bean 对象就可以被直接获取。

按照上述设计过程，我们来实现一个简单的 Spring Bean 容器。

1.3　简单容器实现

1. 工程结构

```
spring-step-01
└── src
    ├── main
    │   └── java
    │       └── cn.bugstack.springframework
    │           ├── BeanDefinition.java
    │           └── BeanFactory.java
    └── test
        └── java
            └── cn.bugstack.springframework.test
                ├── bean
                │   └── UserService.java
                └── ApiTest.java
```

9

> 注意：请按本书引言中介绍的方式获取源码。

Spring Bean 容器中类的关系如图 1-2 所示。

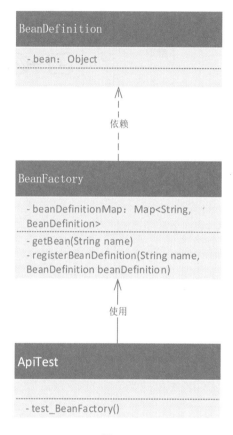

图 1-2

Spring Bean 容器实现的内容非常简单，仅包括一个简单的 BeanFactory 类和 BeanDefinition 类。这里的类名称与 Spring 源码中的类名称一致，只是实现起来会相对简化一些。在后续的扩展过程中，我们可以不断地添加内容。

- BeanDefinition：用于定义 Bean 对象，它的实现方式是以一个 Object 类型存储对象。
- BeanFactory：代表 Bean 对象的工厂，可以将 Bean 对象的定义存储到 Map 中以便获取 Bean 对象。

2. Bean 对象的定义

源码详见：cn.bugstack.springframework.BeanDefinition。

```java
public class BeanDefinition {

    private Object bean;

    public BeanDefinition(Object bean) {
        this.bean = bean;
    }

    public Object getBean() {
        return bean;
    }

}
```

在目前的 Bean 对象定义中，只有一个 Object 类型，因为它是所有类型的父类，可以存储任意类型的 Bean 对象。读者可以参考 Spring 源码中这个类的信息，其名称都是一样的。

在后续章节的实现过程中，我们会逐步完善 BeanDefinition 类对 Bean 对象定义的相关属性字段，如 propertyValues、initMethodName、destroyMethodName、scope、beanClass 等。

3. Bean 工厂

源码详见：cn.bugstack.springframework.BeanFactory。

```java
public class BeanFactory {

    private Map<String, BeanDefinition> beanDefinitionMap = new ConcurrentHashMap<>();

    public Object getBean(String name) {
        return beanDefinitionMap.get(name).getBean();
    }

    public void registerBeanDefinition(String name, BeanDefinition beanDefinition) {
        beanDefinitionMap.put(name, beanDefinition);
    }

}
```

BeanFactory 类是用于生成和使用对象的 Bean 工厂，BeanFactory 类的实现过程包括

Bean 对象的注册和获取，这里注册的是 Bean 对象的定义信息。

目前，BeanFactory 类的实现是非常简化的，但这种简化的实现却是整个 Spring Bean 容器中关于 Bean 对象使用的最终体现，只不过在实现过程中只展示出了基本的核心原理。在后续的补充实现中，这个类的内容会不断增加。

1.4 容器使用测试

1. 事先准备

源码详见：cn.bugstack.springframework.test.bean.UserService。

```
public class UserService {

    public void queryUserInfo(){
        System.out.println(" 查询用户信息 ");
    }

}
```

这里简单定义了一个 UserService 对象类，方便后续对 Spring Bean 容器进行测试。

2. 测试实例

源码详见：cn.bugstack.springframework.test.ApiTest。

```
@Test
public void test_BeanFactory(){
    // 1. 初始化 BeanFactory 接口
    BeanFactory beanFactory = new BeanFactory();

    // 2. 注册 Bean 对象
    BeanDefinition beanDefinition = new BeanDefinition(new UserService());
    beanFactory.registerBeanDefinition("userService", beanDefinition);

    // 3. 获取 Bean 对象
    UserService userService = (UserService) beanFactory.getBean("userService");
    userService.queryUserInfo();
}
```

在单元测试中，主要包括初始化 BeanFactory 接口、注册 Bean 对象、获取 Bean 对象 3 个步骤，在使用效果上贴近于 Spring 框架，但这里会更加简化一些。

在注册 Bean 对象的过程中,这里直接把 UserService 类实例化后作为入参信息传递给 BeanDefinition。在后续的代码中,我们会将这部分内容放入 Bean 工厂中实现。

3. 测试结果

```
查询用户信息
Process finished with exit code 0
```

从测试结果中可以看到,目前的 Spring Bean 容器案例已经有了雏形。

1.5 本章总结

关于 Spring Bean 容器的一个简单实现已经完成了,这部分代码相对简单,读者稍加尝试就可以实现这部分内容。

但对于学习知识的过程来说,写代码只是最后的步骤,形成整体思路和设计方案才是更重要的。读者只有厘清了思路,才能真正地理解代码。

第 2 章会在此工程的基础上扩展实现,要比本章介绍的类多一些。同时,每一个功能的实现都会基于一个具体的需求进行目标分析和方案设计,以便于读者在学习编码的同时更加注重技术价值的学习。

第 2 章 实现 Bean 对象的定义、注册和获取

初学者通常很难读懂 Spring 源码，不仅是因为 Spring 框架变得越来越庞大和复杂，更是因为在 Spring 源码的实现过程中融入了很多设计模式的原则和使用方法，让 Spring 的功能更具有可扩展性和可维护性。也正因如此，Spring 才有了更顽强的生命力。

因为大部分研发人员每天只与简单的业务功能迭代接触，缺少运用设计模式解决复杂业务场景的开发经验，所以很难清楚地阅读框架源码。

- 本章难度：★★☆☆☆
- 本章重点：我们通过使用 AbstractBeanFactory 抽象类，运用模板模式拆分功能，解耦 Spring Bean 容器，处理界限上下文关系，完成 BeanFactory 接口的实现。本章的代码量并不多，但设计模式在 Spring 框架中的开发运用可在本章中初见雏形。这样的设计可以让代码更加符合高内聚、低耦合的设计理念。

2.1 容器的思考

第 1 章依照 Spring Bean 容器的概念，初步实现了简单版本的容器代码。本章需要结合已经实现的 Spring Bean 容器功能，实现 Spring Bean 容器关于 Bean 对象的定义、注册和获取功能。

首先，应该通过 Spring Bean 容器创建 Bean 对象，而不是在调用时传递一个完成了实例化的 Bean 对象。然后，还需要考虑单例对象，在二次获取对象时，可以从内存中获取。

此外，不仅要实现功能，还需要完善基础容器框架的类结构体，否则很难将其他的功能添加进去。下面使用设计模式来开发 Spring Bean 容器的功能结构，按照类的功能拆分出不同的接口和实现类。

2.2 完善容器设计

鉴于本节的需求，需要先完善 Spring Bean 容器。在注册 Bean 对象时，只注册一个类信息，而不直接将实例化信息注册到 Spring Bean 容器中。这里首先需要将 BeanDefinition 类中的 Object 属性修改为 Class，然后在获取 Bean 对象时进行 Bean 对象的实例化，以及判断当前单例对象在容器中是否已经被缓存，如图 2-1 所示。

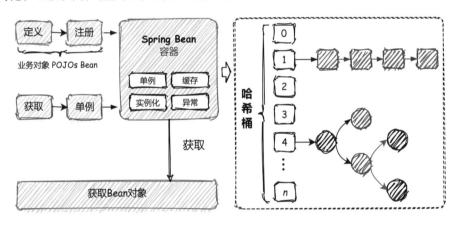

图 2-1

首先需要定义像 BeanFactory 这样的 Bean 工厂，提供 Bean 对象的获取方法 getBean(String name)，然后使用 AbstractBeanFactory 抽象类实现 BeanFactory 接口。我们通过使用模板模式，可以统一使用核心方法的调用逻辑和标准定义，进而很好地实现后续的步骤而不用关心某一方法调用逻辑。按照统一方式执行，类的继承者只需要关心具体方法如何实现即可。

继承 AbstractBeanFactory 抽象类后的 AbstractAutowireCapableBeanFactory 类可以实现

相应的抽象方法，因为 AbstractAutowireCapableBeanFactory 本身也是一个抽象类。但它只实现属于自己的抽象方法即可，其他抽象方法由继承 AbstractAutowireCapableBeanFactory 的类来实现。这里体现了在类的实现过程中各司其职的特点。

另外，还有非常重要的一点——关于单例对象 SingletonBeanRegistry 的接口定义。DefaultSingletonBeanRegistry 接口实现后，会被 AbstractBeanFactory 抽象类继承。此时，AbstractBeanFactory 是一个非常完整和强大的抽象类，可以非常好地体现出它对模板模式的抽象定义。

2.3 完善容器实现

1. 工程结构

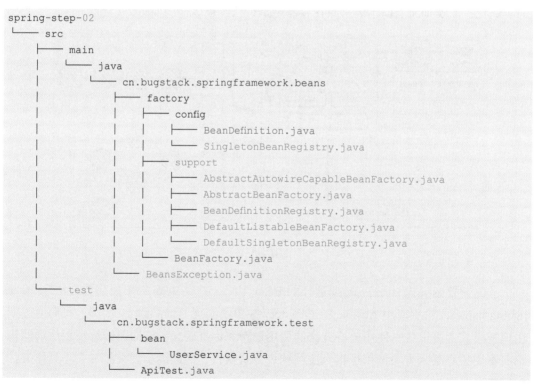

Spring Bean 容器中类的关系如图 2-2 所示。

第 2 章 实现 Bean 对象的定义、注册和获取

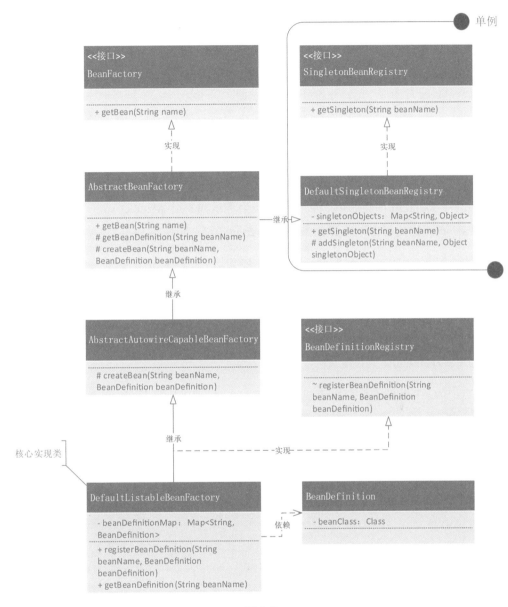

图 2-2

虽然本节关于 Spring Bean 容器功能的实现代码与 Spring 源码还有不小的差距，但从实现结果的类关系图来看，其实已经具备了一定的设计复杂性。这些复杂的类关系在各个接口的定义和实现及抽象类的继承中都有所体现，具体如下。

- BeanFactory 接口是通过 AbstractBeanFactory 抽象类实现的 getBean 方法来定义的。
- AbstractBeanFactory 抽象类继承了 SingletonBeanRegistry 接口的 DefaultSingletonBeanRegistry 类。所以，AbstractBeanFactory 抽象类就具备了单例 Bean 对象的注册功能。
- AbstractBeanFactory 定义了两个抽象方法——getBeanDefinition(String beanName) 和 createBean(String beanName, BeanDefinition beanDefinition)，它们分别由 DefaultListableBeanFactory 类和 AbstractAutowireCapableBeanFactory 类实现。
- 最终，DefaultListableBeanFactory 类还会继承抽象类 AbstractAutowireCapableBeanFactory，因此可以调用该抽象类中的 createBean 方法。

这部分类的关系和实现过程有一些复杂，因为所有的实现都以职责划分、共性分离，以调用关系定义为标准搭建的类关系。

> 注意：这部分内容可以拓展研发人员在复杂业务系统开发中的设计思路。

2. BeanDefinition 类的定义

源码详见：cn.bugstack.springframework.beans.factory.config.BeanDefinition。

```java
public class BeanDefinition {

    private Class beanClass;

    public BeanDefinition(Class beanClass) {
        this.beanClass = beanClass;
    }

    // ...get/set
}
```

在 Bean 对象定义类中，将 2.3 节中 Bean 对象定义源码中的 Object bean 替换为 Class，这样就可以将 Bean 对象的实例化放到容器中处理。

> 注意：如果仔细阅读第 1 章并做了相应的测试，就会发现 Bean 对象的实例化是在初始化调用阶段由 BeanDefinition 构造函数完成的。

3. 单例对象注册接口的定义和实现

源码详见：cn.bugstack.springframework.beans.factory.config.SingletonBeanRegistry。

```
public interface SingletonBeanRegistry {

    Object getSingleton(String beanName);

    void registerSingleton(String beanName, Object singletonObject);

}
```

SingletonBeanRegistry 类比较简单，主要用于定义一个注册和获取单例对象的接口。

源码详见：cn.bugstack.springframework.beans.factory.config.DefaultSingletonBeanRegistry。

```
public class DefaultSingletonBeanRegistry implements SingletonBeanRegistry {

    private Map<String, Object> singletonObjects = new HashMap<>();

    @Override
    public Object getSingleton(String beanName) {
        return singletonObjects.get(beanName);
    }

    @Override
    public void registerSingleton(String beanName, Object singletonObject) {
        singletonObjects.put(beanName, singletonObject);
    }

}
```

DefaultSingletonBeanRegistry 类主要实现获取单例对象的方法（getSingleton）和注册单例对象的方法（registerSingleton），这两个方法都可以被继承此类的其他类调用，如 AbstractBeanFactory 类及 DefaultListableBeanFactory 类。

4. 抽象类定义模板方法（AbstractBeanFactory）

源码详见：cn.bugstack.springframework.beans.factory.support.AbstractBeanFactory。

```
public abstract class AbstractBeanFactory extends DefaultSingletonBeanRegistry
implements BeanFactory {

    @Override
    public Object getBean(String name) throws BeansException {
```

```
        Object bean = getSingleton(name);
        if (bean != null) {
            return bean;
        }

        BeanDefinition beanDefinition = getBeanDefinition(name);
        return createBean(name, beanDefinition);
    }

    protected abstract BeanDefinition getBeanDefinition(String beanName) throws BeansException;

    protected abstract Object createBean(String beanName, BeanDefinition beanDefinition)
throws BeansException;

}
```

上述代码运用模板模式定义了一个流程标准的用于获取对象的 AbstractBeanFactory 抽象类，并采用职责分离的结构设计，继承 DefaultSingletonBeanRegistry 类，使用其提供的单例对象注册和获取功能，通过 BeanFactory 接口提供一个功能单一的方法，屏蔽了内部逻辑细节。

这里的 BeanFactory 接口提供了一个获取对象的方法 getBean，然后由抽象类实现细节。在 getBean 方法中可以看到，它主要用于获取单例 Bean 对象，以及在无法获取 Bean 对象时做相应的 Bean 对象实例化。getBean 自身并没有实现这些方法，只是定义了调用过程并提供了抽象方法，由此实现抽象类中其他方法的相应功能。

继承 AbstractBeanFactory 抽象类的类有两个，包括 AbstractAutowireCapableBeanFactory 类和 DefaultListableBeanFactory 类，源码中对这两个类也进行了相应的实现。

5. 实例化 Bean 对象（AbstractAutowireCapableBeanFactory）

源码详见：cn.bugstack.springframework.beans.factory.support.AbstractAutowireCapableBeanFactory。

```
public abstract class AbstractAutowireCapableBeanFactory extends AbstractBeanFactory {

    @Override
    protected Object createBean(String beanName, BeanDefinition beanDefinition) throws
BeansException {
        Object bean = null;
        try {
            bean = beanDefinition.getBeanClass().newInstance();
```

```
        } catch (InstantiationException | IllegalAccessException e) {
            throw new BeansException("Instantiation of bean failed", e);
        }

        registerSingleton(beanName, bean);
        return bean;
    }
}
```

继续细分功能职责，AbstractAutowireCapableBeanFactory 抽象类继承 AbstractBeanFactory 类，用于实现创建对象的具体功能。因为它是一个抽象类，所以可以只实现其中部分抽象类接口。另外，用于实现 Bean 对象实例化的 newInstance 方法中存在一个问题：如果对象中含有带入参信息的构造函数，那么该如何处理？

在完成 Bean 对象的实例化后，可以直接调用 registerSingleton 方法，将单例对象存储到缓存中。

6. 核心类实现（DefaultListableBeanFactory）

源码详见：cn.bugstack.springframework.beans.factory.support.DefaultListableBeanFactory。

```
public class DefaultListableBeanFactory extends AbstractAutowireCapableBeanFactory
implements BeanDefinitionRegistry {

    private Map<String, BeanDefinition> beanDefinitionMap = new HashMap<>();

    @Override
    public void registerBeanDefinition(String beanName, BeanDefinition beanDefinition) {
        beanDefinitionMap.put(beanName, beanDefinition);
    }

    @Override
    public BeanDefinition getBeanDefinition(String beanName) throws BeansException {
        BeanDefinition beanDefinition = beanDefinitionMap.get(beanName);
        if (beanDefinition == null) throw new BeansException("No bean named '" + beanName + "' is defined");
        return beanDefinition;
    }

}
```

DefaultListableBeanFactory 在 Spring 源码中也是一个非常核心的类，与源码中的类名保持一致。

当 DefaultListableBeanFactory 类继承了 AbstractAutowireCapableBeanFactory 类后，也就具有了 BeanFactory 类和 AbstractBeanFactory 类的功能。

> 注意：有时会看到一些强制转换的类调用了某些方法，是因为强制转换的类实现了接口功能或继承了某些类。

除此之外，DefaultListableBeanFactory 类还实现了 BeanDefinitionRegistry 类中的 registerBeanDefinition (String beanName, BeanDefinition beanDefinition) 方法，当然还会看到 getBeanDefinition 方法的实现，这个方法在前文中提到过，它是在 AbstractBeanFactory 抽象类中定义的抽象方法。

2.4 Bean 生命周期测试

1. 事先准备

源码详见：cn.bugstack.springframework.test.bean.UserService。

```
public class UserService {

    public void queryUserInfo(){
        System.out.println(" 查询用户信息 ");
    }

}
```

这里简单地定义了一个 UserService 对象，方便后续对 Spring Bean 容器进行测试。

2. 测试实例

源码详见：cn.bugstack.springframework.test.ApiTest。

```
@Test
public void test_BeanFactory(){
    // 1. 初始化 BeanFactory 接口
    DefaultListableBeanFactory beanFactory = new DefaultListableBeanFactory();

    // 2. 注册 Bean 对象
    BeanDefinition beanDefinition = new BeanDefinition(UserService.class);
    beanFactory.registerBeanDefinition("userService", beanDefinition);
```

```
    // 3. 获取 Bean 对象
    UserService userService = (UserService) beanFactory.getBean("userService");
    userService.queryUserInfo();

    // 4. 再次获取和调用 Bean 对象
    UserService userService_singleton = (UserService) beanFactory.getBean("userService");
    userService_singleton.queryUserInfo();
}
```

在此次的单元测试中，除了包括初始化 BeanFactory 接口、注册 Bean 对象、获取 Bean 对象 3 个步骤，还增加了一次对 Bean 对象的获取和调用。这里主要测试和验证单例对象是否能被正确地存储到缓存中。

此外，与第 1 章测试过程不同的是，我们将 UserService.class 传递给了 BeanDefinition，而不是直接执行 new UserService 操作。

3. 测试结果

```
查询用户信息
查询用户信息

Process finished with exit code 0
```

这里显示了两次测试结果，一次是获取 Bean 对象时直接创建的对象，另一次是从缓存中获取的实例化对象。

从调试的截图中也可以看到第二次获取的单例对象，说明已经可以从内存中获取这些单例对象了，如图 2-3 所示。

图 2-3

本章的功能实现和测试验证已经完成。对于不熟悉的内容，读者可以通过设置断点来调试各个阶段类的调用功能，熟悉调用关系。

2.5 本章总结

相对于第 1 章对 Spring Bean 容器的简单实现，本章完善了功能。在实现的过程中，也可以看到类的关系变得越来越复杂了。如果没有做过一些稍微复杂的系统设计开发，那么在利用类定义容器时可能有些困难。

对于 Spring Bean 容器的实现类，要重点关注类之间的职责和关系。几乎所有程序的功能设计都离不开接口、抽象类的实现和继承。使用这些特性不同的类，可以非常好地将类的功能职责和作用范围区分开。这些也是学习开发 Spring Bean 容器框架过程中非常重要的知识。

此外，在编写程序时，要尽量设想未来系统功能迭代的复杂性，从而思考功能逻辑的职责边界、系统服务的界限上下文、合理拆解模块的实现分层，并运用设计模式做好代码的开发设计。

第 3 章
基于 Cglib 实现含构造函数的类实例化策略

本章采用渐进式开发方式，实现 Spring 框架的核心功能。在这个过程中，我们会带领读者共同思考当前的内容还有哪些需要完善，以及应该从哪些角度出发去设计和实现这些功能。

对于研发人员来说，无论是编写业务需求、技术组件，还是编写一个简单的 Spring 框架，重要的前提都是思考和设计。编码只是确定方案后具体行为逻辑的落地过程，研发人员只有不断地琢磨细节、推敲方案、反复验证，才能不断成长。

- 本章难度：★★☆☆☆
- 本章重点：基于策略模式实现两种用于实例化对象的方法，如 JDK、Cglib，并基于此方法实现含有带入参信息的构造函数的类实例化策略。

3.1 实例化问题

第 2 章详细介绍了 Spring Bean 容器的功能，可以把实例化对象交给容器进行统一处理。但是在实例化对象的代码中，并没有考虑类中是否含有带入参信息的构造函数。也就是说，如果实例化一个含有带入参信息的构造函数的类，就会抛出异常。

怎么验证类中是否含有带入参信息的构造函数呢？我们可以在 UserService 的 Bean

对象中添加一个含有带入参信息的构造函数的类，代码如下：

```
public class UserService {

    private String name;

    public UserService(String name) {
        this.name = name;
    }

    // ...
}
```

上述代码报错如下：

```
java.lang.InstantiationException: cn.bugstack.springframework.test.bean.UserService

    at java.lang.Class.newInstance(Class.java:427)
    at cn.bugstack.springframework.test.ApiTest.test_newInstance(ApiTest.java:51)
    ...
```

发生这个问题的主要原因是 beanDefinition.getBeanClass().newInstance() 的实例化方法并没有考虑类中含有带入参信息的构造函数，所以出现实例化异常。

3.2 实例化策略设计

解决上述问题主要考虑两方面，一方面是如何将构造函数的入参信息合理地传递到实例化中，另一方面是如何将含有带入参信息的构造函数的类实例化，如图 3-1 所示。

参考 Spring Bean 容器源码的实现方式，在 BeanFactory 类中添加 Object getBean (String name, Object... args) 接口，这样就可以在获取 Bean 对象时将构造函数的入参信息传递进去。

那么使用什么方式可以创建包含构造函数的 Bean 对象呢？这里有两种方式可以选择，一种是基于 Java 本身自带的方法 DeclaredConstructor，另一种是使用 Cglib 动态创建 Bean 对象。其中，因为 Cglib 是基于 ASM 字节码框架实现的，所以也可以直接通过 ASM 指令码创建 Bean 对象。

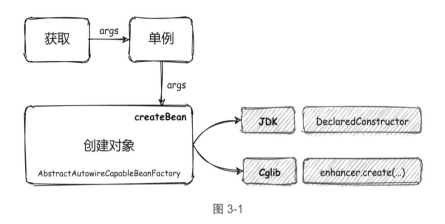

图 3-1

3.3 实例化策略代码实现

1. 工程结构

```
spring-step-03
└── src
    ├── main
    │   └── java
    │       └── cn.bugstack.springframework.beans
    │           ├── factory
    │           │   ├── config
    │           │   │   ├── BeanDefinition.java
    │           │   │   └── SingletonBeanRegistry.java
    │           │   ├── support
    │           │   │   ├── AbstractAutowireCapableBeanFactory.java
    │           │   │   ├── AbstractBeanFactory.java
    │           │   │   ├── BeanDefinitionRegistry.java
    │           │   │   ├── CglibSubclassingInstantiationStrategy.java
    │           │   │   ├── DefaultListableBeanFactory.java
    │           │   │   ├── DefaultSingletonBeanRegistry.java
    │           │   │   ├── InstantiationStrategy.java
    │           │   │   └── SimpleInstantiationStrategy.java
    │           │   └── BeanFactory.java
    │           └── BeansException.java
    └── test
        └── java
```

```
└── cn.bugstack.springframework.test
    ├── bean
    │   └── UserService.java
    └── ApiTest.java
```

Spring Bean 容器中类的关系如图 3-2 所示。

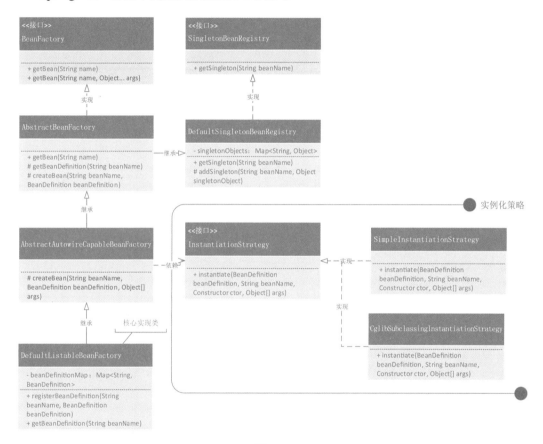

图 3-2

本节的核心是在现有工程中添加 InstantiationStrategy 实例化策略接口、开发对应的 JDK 和 Cglib 实例化方式，以及补充相应的含有带入参信息的 getBean 构造函数。当进行外部调用时，可以传递构造函数的入参信息并进行实例化。

2. 新增 getBean 接口

源码详见：cn.bugstack.springframework.beans.factory.BeanFactory。

```java
public interface BeanFactory {

    Object getBean(String name) throws BeansException;

    Object getBean(String name, Object... args) throws BeansException;

}
```

在 BeanFactory 类中，重载了一个含有带入参信息 args 的 getBean 构造函数，这样就可以将构造函数的入参信息传递给实例化方法进行实例化。

3. 定义实例化策略接口

源码详见：cn.bugstack.springframework.beans.factory.support.InstantiationStrategy。

```java
public interface InstantiationStrategy {

    Object instantiate(BeanDefinition beanDefinition, String beanName, Constructor ctor, Object[] args) throws BeansException;

}
```

在 instantiate 构造函数中添加必要的入参信息，包括 beanDefinition、beanName、ctor 和 args。

读者对于 Constructor 可能会有一点陌生。它是 java.lang.reflect 包下的类，包含了一些必要的类信息。参数 ctor 的作用是获取与入参信息相对应的构造函数。

而 args 是一个具体的入参信息，在实例化对象时会使用到。

4. JDK 实例化

源码详见：cn.bugstack.springframework.beans.factory.support.SimpleInstantiationStrategy。

```java
public class SimpleInstantiationStrategy implements InstantiationStrategy {

    @Override
    public Object instantiate(BeanDefinition beanDefinition, String beanName, Constructor ctor, Object[] args) throws BeansException {
        Class clazz = beanDefinition.getBeanClass();
        try {
            if (null != ctor) {
                return clazz.getDeclaredConstructor(ctor.getParameterTypes()).newInstance(args);
            } else {
```

```
                return clazz.getDeclaredConstructor().newInstance();
            }
        } catch (NoSuchMethodException | InstantiationException | IllegalAccessException
 | InvocationTargetException e) {
            throw new BeansException("Failed to instantiate [" + clazz.getName() + "]", e);
        }
    }
}
```

首先，通过 beanDefinition 获取 Class 信息。Class 信息是在定义 Bean 对象时传递进去的。

然后，判断 ctor 是否为空，如果为空，则无构造函数可实例化；如果不为空，则需要传递构造函数的入参信息进行实例化，实例化方法为 clazz.getDeclaredConstructor (ctor.getParameterTypes()).newInstance(args)，即将入参信息传递给 newInstance 进行实例化。

5. Cglib 实例化

源码详见：cn.bugstack.springframework.beans.factory.support.CglibSubclassingInstantiationStrategy。

```
public class CglibSubclassingInstantiationStrategy implements InstantiationStrategy {

    @Override
    public Object instantiate(BeanDefinition beanDefinition, String beanName, Constructor ctor, Object[] args) throws BeansException {
        Enhancer enhancer = new Enhancer();
        enhancer.setSuperclass(beanDefinition.getBeanClass());
        enhancer.setCallback(new NoOp() {
            @Override
            public int hashCode() {
                return super.hashCode();
            }
        });
        if (null == ctor) return enhancer.create();
        return enhancer.create(ctor.getParameterTypes(), args);
    }

}
```

使用 Cglib 创建含有构造函数的 Bean 对象也非常方便，这里进行了简化处理。在阅读 Spring 源码时，还会学习到 CallbackFilter 等实现方式。

6. 创建策略调用

源码详见：cn.bugstack.springframework.beans.factory.support.AbstractAutowireCapableBeanFactory。

```java
public abstract class AbstractAutowireCapableBeanFactory extends AbstractBeanFactory {

    private InstantiationStrategy instantiationStrategy = new CglibSubclassingInstantiationStrategy();

    @Override
    protected Object createBean(String beanName, BeanDefinition beanDefinition, Object[] args) throws BeansException {
        Object bean = null;
        try {
            bean = createBeanInstance(beanDefinition, beanName, args);
        } catch (Exception e) {
            throw new BeansException("Instantiation of bean failed", e);
        }

        registerSingleton(beanName, bean);
        return bean;
    }

    protected Object createBeanInstance(BeanDefinition beanDefinition, String beanName, Object[] args) {
        Constructor constructorToUse = null;
        Class<?> beanClass = beanDefinition.getBeanClass();
        Constructor<?>[] declaredConstructors = beanClass.getDeclaredConstructors();
        for (Constructor ctor : declaredConstructors) {
            if (null != args && ctor.getParameterTypes().length == args.length) {
                constructorToUse = ctor;
                break;
            }
        }
        return getInstantiationStrategy().instantiate(beanDefinition, beanName, constructorToUse, args);
    }

}
```

首先，在 AbstractAutowireCapableBeanFactory 抽象类中定义了一个对象实例化策略属性类 InstantiationStrategy instantiationStrategy，这里使用 Cglib 类实现。

然后，使用 createBeanInstance 方法，在这个方法中需要注意 Constructor 表示有多少个构造函数，通过 beanClass.getDeclaredConstructors 方法可以获取所有的构造函数，形成一个集合。

最后，需要循环对比构造函数集合与入参信息 args 的匹配情况。这里对比的方式比较简单，只通过数量对比即可，而在实际的 Spring 源码中，还需要对比入参类型，否则在遇到数量相同、入参类型不同时，就会抛出异常。读者完成这部分核心链路的学习后，可以结合 Spring 的源码进行扩展练习。

3.4 构造函数对象测试

1. 事先准备

源码详见：cn.bugstack.springframework.test.bean.UserService。

```
public class UserService {

    private String name;

    public UserService(String name) {
        this.name = name;
    }

    public void queryUserInfo() {
        System.out.println("查询用户信息: " + name);
    }

    @Override
    public String toString() {
        final StringBuilder sb = new StringBuilder("");
        sb.append("").append(name);
        return sb.toString();
    }
}
```

这里在 UserService 中唯一多添加的是一个含有带 name 入参信息的构造函数，目的是方便验证类是否能被实例化。

2. 测试实例

源码详见：cn.bugstack.springframework.test.ApiTest。

```
@Test
public void test_BeanFactory() {
    // 1. 初始化 BeanFactory 接口
    DefaultListableBeanFactory beanFactory = new DefaultListableBeanFactory();

    // 2. 注册 Bean 对象
    BeanDefinition beanDefinition = new BeanDefinition(UserService.class);
    beanFactory.registerBeanDefinition("userService", beanDefinition);

    // 3. 获取 Bean 对象
    UserService userService = (UserService) beanFactory.getBean("userService", "小傅哥");
    userService.queryUserInfo();
}
```

在此次的单元测试中，依然包括 3 个核心步骤：初始化 BeanFactory 接口、注册 Bean 对象、获取 Bean 对象。与第 2 章不同的是，在获取 Bean 对象时传递了一个参数名为"小傅哥"的入参信息，这个信息的传递将会帮助我们创建出含有 String 类型构造函数的 UserService 类，而不会再出现初始化报错的问题。

3. 测试结果

```
查询用户信息：小傅哥

Process finished with exit code 0
```

从测试结果中可以看到，最大的不同是将含有带入参信息的构造函数的类实例化了。

此外，在测试时可以分别尝试使用两种不同的实例化策略：SimpleInstantiationStrategy 和 CglibSubclassingInstantiationStrategy，验证测试结果。

4. 操作案例

这里将几种不同的实例化方式放到单元测试中，方便读者对比理解。

（1）验证无构造函数实例化。

```
@Test
public void test_newInstance() throws IllegalAccessException, InstantiationException {
    UserService userService = UserService.class.newInstance();
    System.out.println(userService);
}
```

这种实例化方式也是在第 2 章实现 Spring Bean 容器时使用的方式。

（2）验证有构造函数的类实例化。

```
@Test
public void test_constructor() throws Exception {
    Class<UserService> userServiceClass = UserService.class;
    Constructor<UserService> declaredConstructor = userServiceClass.getDeclaredConstructor
(String.class);
    UserService userService = declaredConstructor.newInstance(" 小傅哥 ");
    System.out.println(userService);
}
```

如果有构造函数的类需要实例化，则要先使用 getDeclaredConstructor 获取构造函数，再传递入参信息进行实例化。

（3）获取构造函数。

```
@Test
public void test_parameterTypes() throws Exception {
    Class<UserService> beanClass = UserService.class;
    Constructor<?>[] declaredConstructors = beanClass.getDeclaredConstructors();
    Constructor<?> constructor = declaredConstructors[0];
    Constructor<UserService> declaredConstructor = beanClass.getDeclaredConstructor
(constructor.getParameterTypes());
    UserService userService = declaredConstructor.newInstance(" 小傅哥 ");
    System.out.println(userService);
}
```

这个实例最核心的地方在于获取一个类中所有的构造函数，可以使用 beanClass.getDeclaredConstructors 方法实现。

（4）Cglib 实例化。

```
@Test
public void test_cglib() {
    Enhancer enhancer = new Enhancer();
    enhancer.setSuperclass(UserService.class);
    enhancer.setCallback(new NoOp() {
        @Override
        public int hashCode() {
            return super.hashCode();
        }
    });
    Object obj = enhancer.create(new Class[]{String.class}, new Object[]{" 小傅哥 "});
    System.out.println(obj);
}
```

Cglib 在 Spring Bean 容器中的使用方式非常多，有兴趣的读者可以深入学习 Cglib 的扩展内容，以及 ASM 相关技术。

3.5 本章总结

本章以完善实例化，添加 InstantiationStrategy 实例化策略，以及新增两个实例化类为主要内容。这部分涉及的类可以从 Spring 源码中找到对应的实现代码。

从不断完善、增加需求的过程中可以看出，当代码结构设计得较为合理时，就可以容易且方便地扩展不同属性类的职责，而不会因为需求的增加导致类的结构混乱。所以，在实现业务需求的过程中，读者也要尽可能地考虑如何保证良好的扩展性及拆分类的职责。

第 4 章 注入属性和依赖对象

经过前面几章的实践，我们已经可以看到一个初具雏形的 Spring 框架，并且在编写代码时也使用了很多设计模式的技巧，相信读者在实践过程中一定有所体会。

本章继续扩展 Bean 对象在实例化过程中需要补充的内容，这是对 Bean 对象生产过程的细分拆解，只有这样才能让 Spring Bean 容器管理 Bean 对象的整个生命周期。

- 本章难度：★★☆☆☆
- 本章重点：通过模板模式细分拆解 Bean 对象的实例化过程，在第 3 章创建 Bean 对象的基础上，将 PropertyValues 引入 BeanDefinition 定义中，在对象实例化完成后，填充 Bean 对象的属性。

4.1 Bean 对象拆解思考

回顾前面章节的内容，我们介绍了实现容器、定义和注册 Bean 对象、将 Bean 对象实例化、按照是否包含带入参的构造函数实现不同的实例化策略。但是，在创建对象的实例化细分拆解过程中，其实我们还没有思考关于"类中是否有属性"的问题，如果类中包含属性，那么在实例化时就需要填充属性信息，这样才能创建一个完整的对象。

创建对象过程所需填充的属性不只是 int、long、double 等基本类型，还包括可能没有被实例化的对象属性，这些都需要在创建 Bean 对象时填充。不过这里暂时不会考虑 Bean 对象的循环依赖，否则会把整个功能的实现范围扩大，对系统的渐进式迭代开发不够友好。当我们实现核心功能后，再逐步完善循环依赖。

4.2 属性填充设计

因为属性填充是在使用 newInstance 或者 Cglib 创建 Bean 对象后开始执行的，所以可以在 AbstractAutowireCapableBeanFactory 类的 createBean 方法中添加属性填充操作 applyPropertyValues，如图 4-1 所示。

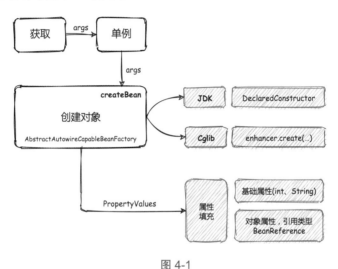

图 4-1

> 注意：这部分内容也可以对照 Spring 源码学习，本节是 Spring 框架的简化版实现，对照学习会更易于理解。

由于在创建 Bean 对象时要执行属性填充操作，所以要在定义 Bean 对象的 BeanDefinition 类中添加对象创建时所需要的 PropertyValues 属性集合。

填充的信息还包括 Bean 对象的类型，即需要再定义一个 BeanReference 引用对象（相当于借壳），里面只是一个简单的 Bean 对象名称，在具体实例化时进行递归创建和填充，与 Spring 源码中的实现一样。

> 注意：在 Spring 源码中，BeanReference 是一个接口。

4.3　属性填充实现

1. 工程结构

```
spring-step-04
└── src
    ├── main
    │   └── java
    │       └── cn.bugstack.springframework.beans
    │           ├── factory
    │           │   ├── config
    │           │   │   ├── BeanDefinition.java
    │           │   │   ├── BeanReference.java
    │           │   │   └── SingletonBeanRegistry.java
    │           │   ├── support
    │           │   │   ├── AbstractAutowireCapableBeanFactory.java
    │           │   │   ├── AbstractBeanFactory.java
    │           │   │   ├── BeanDefinitionRegistry.java
    │           │   │   ├── CglibSubclassingInstantiationStrategy.java
    │           │   │   ├── DefaultListableBeanFactory.java
    │           │   │   ├── DefaultSingletonBeanRegistry.java
    │           │   │   ├── InstantiationStrategy.java
    │           │   │   └── SimpleInstantiationStrategy.java
    │           │   └── BeanFactory.java
    │           ├── BeansException.java
    │           ├── PropertyValue.java
    │           └── PropertyValues.java
    └── test
        └── java
            └── cn.bugstack.springframework.test
                ├── bean
                │   ├── UserDao.java
                │   └── UserService.java
                └── ApiTest.java
```

Spring Bean 容器中类的关系如图 4-2 所示。

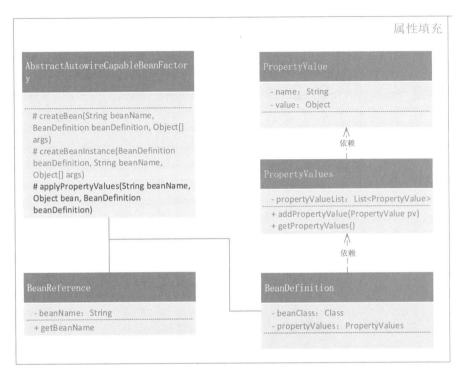

图 4-2

本节需要增加 3 个类：BeanReference（类引用）、PropertyValue（属性值）、PropertyValues（属性集合），分别用于类和其他类型属性的填充。

另外，需要改动的主要是 AbstractAutowireCapableBeanFactory 类的 createBean 方法中用于属性填充的部分。

2．定义属性

源码详见：cn.bugstack.springframework.beans.PropertyValue。

```
public class PropertyValue {

    private final String name;

    private final Object value;

    public PropertyValue(String name, Object value) {
        this.name = name;
        this.value = value;
```

```
    }
    
    // ...get/set
}
```

源码详见：cn.bugstack.springframework.beans.PropertyValues。

```java
public class PropertyValues {

    private final List<PropertyValue> propertyValueList = new ArrayList<>();

    public void addPropertyValue(PropertyValue pv) {
        this.propertyValueList.add(pv);
    }

    public PropertyValue[] getPropertyValues() {
        return this.propertyValueList.toArray(new PropertyValue[0]);
    }

    public PropertyValue getPropertyValue(String propertyName) {
        for (PropertyValue pv : this.propertyValueList) {
            if (pv.getName().equals(propertyName)) {
                return pv;
            }
        }
        return null;
    }

}
```

PropertyValue 和 PropertyValues 这两个类的作用就是传递 Bean 对象创建过程中所需要的属性信息，因为可能会有很多个属性，所以需要定义一个 PropertyValues 类进行属性集合包装。

3. 补全 Bean 对象定义

源码详见：cn.bugstack.springframework.beans.factory.config.BeanDefinition。

```java
public class BeanDefinition {

    private Class beanClass;

    private PropertyValues propertyValues;

    public BeanDefinition(Class beanClass) {
```

```
        this.beanClass = beanClass;
        this.propertyValues = new PropertyValues();
    }

    public BeanDefinition(Class beanClass, PropertyValues propertyValues) {
        this.beanClass = beanClass;
        this.propertyValues = propertyValues != null ? propertyValues : new PropertyValues();
    }

    // ...get/set
}
```

在定义 Bean 对象的过程中需要传递 Bean 对象属性信息,这在前面章节的测试中都有所体现,即 new BeanDefinition(UserService.class, propertyValues)。

为了传递属性信息,这里填充了 PropertyValues 属性,同时在 BeanDefinition (Class beanClass, PropertyValues propertyValues) 构造函数中对 PropertyValues 属性进行了判断处理。如果值为空,则进行实例化并创建一个 PropertyValues 类,将其填充到 BeanDefinition propertyValues 属性上。

4. Bean 对象属性填充

源码详见: cn.bugstack.springframework.beans.factory.support.AbstractAutowireCapableBeanFactory。

```
public abstract class AbstractAutowireCapableBeanFactory extends AbstractBeanFactory {

    private InstantiationStrategy instantiationStrategy = new CglibSubclassingInstantiationStrategy();

    @Override
    protected Object createBean(String beanName, BeanDefinition beanDefinition, Object[] args) throws BeansException {
        Object bean = null;
        try {
            bean = createBeanInstance(beanDefinition, beanName, args);
            // 给 Bean 对象填充属性
            applyPropertyValues(beanName, bean, beanDefinition);
        } catch (Exception e) {
            throw new BeansException("Instantiation of bean failed", e);
        }

        registerSingleton(beanName, bean);
        return bean;
```

```java
    }

    protected Object createBeanInstance(BeanDefinition beanDefinition, String beanName, Object[] args) {
        Constructor constructorToUse = null;
        Class<?> beanClass = beanDefinition.getBeanClass();
        Constructor<?>[] declaredConstructors = beanClass.getDeclaredConstructors();
        for (Constructor ctor : declaredConstructors) {
            if (null != args && ctor.getParameterTypes().length == args.length) {
                constructorToUse = ctor;
                break;
            }
        }
        return getInstantiationStrategy().instantiate(beanDefinition, beanName, constructorToUse, args);
    }

    /**
     * Bean 对象属性填充
     */
    protected void applyPropertyValues(String beanName, Object bean, BeanDefinition beanDefinition) {
        try {
            PropertyValues propertyValues = beanDefinition.getPropertyValues();
            for (PropertyValue propertyValue : propertyValues.getPropertyValues()) {

                String name = propertyValue.getName();
                Object value = propertyValue.getValue();

                if (value instanceof BeanReference) {
                    // 例如, A 依赖 B, 获取 B 的实例化对象
                    BeanReference beanReference = (BeanReference) value;
                    value = getBean(beanReference.getBeanName());
                }
                // 属性填充
                BeanUtil.setFieldValue(bean, name, value);
            }
        } catch (Exception e) {
            throw new BeansException("Error setting property values: " + beanName);
        }
    }

    public InstantiationStrategy getInstantiationStrategy() {
        return instantiationStrategy;
```

```
    }

    public void setInstantiationStrategy(InstantiationStrategy instantiationStrategy) {
        this.instantiationStrategy = instantiationStrategy;
    }

}
```

AbstractAutowireCapableBeanFactory 类中的内容稍微有点多，主要包括 3 个方法——createBean、createBeanInstance、applyPropertyValues，这里主要关注在 createBean 方法中调用的 applyPropertyValues 方法。

在 applyPropertyValues 方法中，我们通过获取 beanDefinition.getPropertyValues 来循环执行属性填充。如果遇到 BeanReference 引用类型，则需要通过递归获取 Bean 对象实例，调用 getBean 方法。

当完成 Bean 对象创建和依赖后，则可以通过递归调用的方式来完成属性的填充。这里需要注意，我们并没有处理循环依赖的问题，这部分内容较多，会在后续补充介绍。其中，BeanUtil.setFieldValue(bean, name, value) 是 hutool-all 工具类中提供的属性信息设置方法。如果读者没有使用过工具类，则可以基于 Java 提供的反射，对属性信息进行设置。

4.4 注入属性测试

1. 事先准备

源码详见：cn.bugstack.springframework.test.bean.UserDao。

```
public class UserDao {

    private static Map<String, String> hashMap = new HashMap<>();

    static {
        hashMap.put("10001", " 小傅哥 ");
        hashMap.put("10002", " 八杯水 ");
        hashMap.put("10003", " 阿毛 ");
    }

    public String queryUserName(String uId) {
        return hashMap.get(uId);
```

        }

    }

源码详见：cn.bugstack.springframework.test.bean.UserService。

```java
public class UserService {

    private String uId;

    private UserDao userDao;

    public void queryUserInfo() {
        System.out.println("查询用户信息: " + userDao.queryUserName(uId));
    }

    // ...get/set
}
```

Dao、Service 是开发时经常使用的场景。如果在 UserService 中注册 UserDao，就能体现出 Bean 对象属性的依赖关系。

2. 测试实例

```java
@Test
public void test_BeanFactory() {
    // 1. 初始化 BeanFactory 接口
    DefaultListableBeanFactory beanFactory = new DefaultListableBeanFactory();

    // 2. 注册 UserDao
    beanFactory.registerBeanDefinition("userDao", new BeanDefinition(UserDao.class));

    // 3. 使用 UserService 填充属性（uId、userDao）
    PropertyValues propertyValues = new PropertyValues();
    propertyValues.addPropertyValue(new PropertyValue("uId", "10001"));
    propertyValues.addPropertyValue(new PropertyValue("userDao",new BeanReference("userDao")));

    // 4. 使用 UserService 注册 Bean 对象
    BeanDefinition beanDefinition = new BeanDefinition(UserService.class, propertyValues);
    beanFactory.registerBeanDefinition("userService", beanDefinition);

    // 5. 使用 UserService 获取 Bean 对象
    UserService userService = (UserService) beanFactory.getBean("userService");
    userService.queryUserInfo();

}
```

首先，与直接获取 Bean 对象不同，这次需要先将 UserDao 注册到 Spring Bean 容器中，即 beanFactory.registerBeanDefinition("userDao", new BeanDefinition(UserDao.class))。

然后，进行属性填充，一种操作是填充基本类型属性 new PropertyValue("uId", "10001")，另一种操作是填充对象属性 new PropertyValue("userDao",new BeanReference("userDao"))。

最后一步是使用 UserService 获取 Bean 对象，调用方法即可。

3. 测试结果

查询用户信息：小傅哥

Process finished with exit code 0

从测试结果中可以看到，属性填充已经完成，此时可以调用 Dao 方法，如调用 userDao.queryUserName(uId) 查询用户信息。

下面观察在调试的情况下，UserDao 是否进入了 Bean 对象的属性填充流程中，如图 4-3 所示。

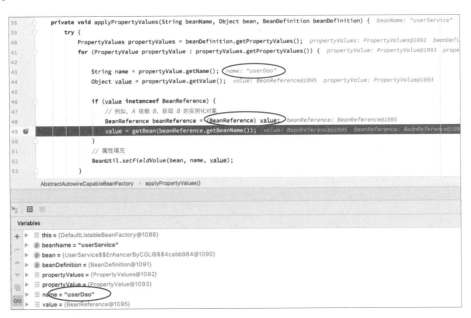

图 4-3

可以看到已经开始进行属性填充了，当属性是 BeanReference 时，需要获取 Bean 对象并创建 Bean 实例。

4.5 本章总结

本章对 AbstractAutowireCapableBeanFactory 类中创建对象的功能进行了扩充，在实现了有构造函数和无构造函数的类实例化策略后，对 Bean 对象的属性信息进行了填充。当 Bean 的属性为 Bean 对象时，需要进行递归处理。最后在进行属性填充时需要用到反射，也可以使用一些工具类。

对于每一章的功能，我们都在循序渐进地实现，这样可以更好地学习 Spring 框架的设计思路，这在对一些已经开发好的类进行新功能扩充时尤为重要。在学习编程时，培养设计思路比仅实现简单的代码更能提升编程能力。

关于 Bean 对象的创建已经完成，接下来需要在整个框架的基础上完成资源属性的加载，这需要对 XML 进行配置。此外，在实现框架的过程中，所有的类名都会参考 Spring 源码，相应的设计实现步骤也与 Spring 源码对应，只是简化了一些流程。读者可以使用相同的类名，在 Spring 源码中检索每一个功能的实现细节并进行补充学习。

第 5 章
资源加载器解析文件注册对象

在逐步开发并实现简单版的 Spring 框架过程中，需要增加新的功能时，我们都是在上一章代码的基础上进行扩展的，但扩展过程更多是使用接口、继承、抽象和封装等特性来实现的，既没有破坏原有的结构，也没有因为增加新的功能产生代码混乱。

通常我们承接的需求功能都是使用 if...else 语句来完成的，但是采用流程化的思路，以及对每一个功能点进行添加、修改、删除，都可能让整个代码逻辑变得松动或者出现一些异常。因此，在学习 Spring 框架开发的同时，更应该学习设计技巧，并将其运用到日常的业务开发中，以提升代码的质量和工程的稳定性。

- 本章难度：★★★☆☆
- 本章重点：定义用于解析 XML 文件的 XmlBeanDefinitionReader 类，处理用户配置在 XML 文件中的 Bean 对象信息，完成自动化配置和注册 Bean 对象。

5.1 对象创建问题

在完成 Spring 框架的雏形后，我们可以通过单元测试，手动完成 Bean 对象的定义、注册和属性填充，以及获取对象调用的方法。但在实际的使用过程中，是不太可能让用户通过手动方式创建对象的，而是要通过 Spring 配置文件来简化创建过程，具体需要完成如下动作。

如图 5-1 所示，首先我们需要将步骤 2、步骤 3、步骤 4 整合到 Spring 框架中，通过

Spring 配置文件将 Bean 对象实例化。

然后在现有的 Spring 框架中添加能实现 Spring 配置的读取、解析和 Bean 对象注册。

```
@Test
public void test_BeanFactory() {
    // 1.初始化 BeanFactory接口
    DefaultListableBeanFactory beanFactory = new DefaultListableBeanFactory();

    // 2. UserDao 注册
    beanFactory.registerBeanDefinition(beanName: "userDao", new BeanDefinition(UserDao.class));

    // 3. 使用UserService 设置属性(uId、userDao)
    PropertyValues propertyValues = new PropertyValues();
    propertyValues.addPropertyValue(new PropertyValue( name: "uId",  value: "10001"));
    propertyValues.addPropertyValue(new PropertyValue( name: "userDao", new BeanReference( beanName: "userDao")));

    // 4. 使用UserService 注册WBean对象
    BeanDefinition beanDefinition = new BeanDefinition(UserService.class, propertyValues);
    beanFactory.registerBeanDefinition( beanName: "userService", beanDefinition);

    // 5. 使用UserService 获取Bean对象
    UserService userService = (UserService) beanFactory.getBean( name: "userService");
    String result = userService.queryUserInfo();
    System.out.println("测试结果: " + result);
}
```

把这一部分操作，放到Spring配置文件中进行处理

图 5-1

5.2 资源加载和解析设计

对于本章的对象创建问题，我们需要在现有的 Spring 框架中添加一个资源加载器，用于读取 ClassPath、本地文件（File）和远程云文件（HTTP 文件）的配置内容。这些配置内容包括 Bean 对象的描述信息和属性信息。当读取配置文件中的内容后，就可以先将配置文件中的 Bean 描述信息解析再注册，将 Bean 对象注册到 Spring Bean 容器中。整体设计结构如图 5-2 所示。

- 资源加载器属于相对独立的部分，位于 Spring 框架核心包下，主要用于读取 ClassPath、本地文件（File）和远程云文件（HTTP 文件）。

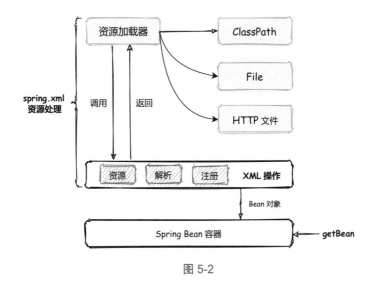

图 5-2

- 当资源被加载后，接下来就是将 Bean 对象解析并注册到 Spring 框架中，这部分需要和 DefaultListableBeanFactory 核心类结合起来实现，因为所有解析后的注册动作，都会将 Bean 对象的定义信息放入 DefaultListableBeanFactory 类中。
- 在实现时，需要设计好接口的层级关系，包括对 Bean 对象的读取接口 BeanDefinitionReader 的定义，以及定义好对应的实现类，在实现类中完成对 Bean 对象的解析和注册。

5.3　资源加载和解析设计实现

1. 工程结构

```
spring-step-05
└── src
    ├── main
    │   └── java
    │       └── cn.bugstack.springframework
    │           └── beans
    │               ├── factory
    │               │   ├── config
```

```
|   |   |   |   |       ├── AutowireCapableBeanFactory.java
|   |   |   |   |       ├── BeanDefinition.java
|   |   |   |   |       ├── BeanReference.java
|   |   |   |   |       ├── ConfigurableBeanFactory.java
|   |   |   |   |       └── SingletonBeanRegistry.java
|   |   |   |   ├── support
|   |   |   |   |   ├── AbstractAutowireCapableBeanFactory.java
|   |   |   |   |   ├── AbstractBeanDefinitionReader.java
|   |   |   |   |   ├── AbstractBeanFactory.java
|   |   |   |   |   ├── BeanDefinitionReader.java
|   |   |   |   |   ├── BeanDefinitionRegistry.java
|   |   |   |   |   ├── CglibSubclassingInstantiationStrategy.java
|   |   |   |   |   ├── DefaultListableBeanFactory.java
|   |   |   |   |   ├── DefaultSingletonBeanRegistry.java
|   |   |   |   |   ├── InstantiationStrategy.java
|   |   |   |   |   └── SimpleInstantiationStrategy.java
|   |   |   |   ├── xml
|   |   |   |   |   └── XmlBeanDefinitionReader.java
|   |   |   |   ├── BeanFactory.java
|   |   |   |   ├── ConfigurableListableBeanFactory.java
|   |   |   |   ├── HierarchicalBeanFactory.java
|   |   |   |   └── ListableBeanFactory.java
|   |   |   ├── BeansException.java
|   |   |   ├── PropertyValue.java
|   |   |   └── PropertyValues.java
|   |   ├── core.io
|   |   |   ├── ClassPathResource.java
|   |   |   ├── DefaultResourceLoader.java
|   |   |   ├── FileSystemResource.java
|   |   |   ├── Resource.java
|   |   |   ├── ResourceLoader.java
|   |   |   └── UrlResource.java
|   |   └── utils
|   |       └── ClassUtils.java
└── test
    └── java
        └── cn.bugstack.springframework.test
            ├── bean
            |   ├── UserDao.java
            |   └── UserService.java
            └── ApiTest.java
```

Spring Bean 容器中资源加载和解析的类关系如图 5-3 所示。

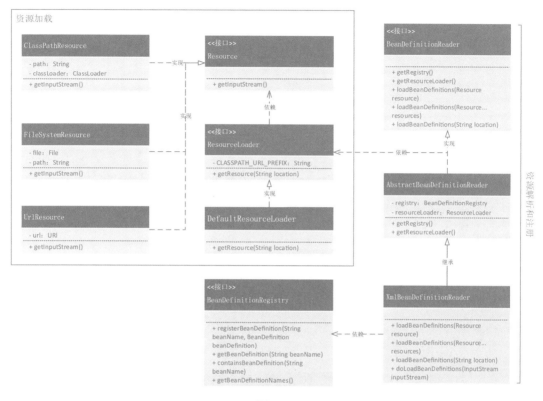

图 5-3

为了将 Bean 对象的定义、注册和初始化交给 spring.xml 配置文件进行处理，需要实现两部分内容——资源加载器、XML 资源处理类。其实现过程主要是对 Resource 接口、ResourceLoader 接口的实现，此外，BeanDefinitionReader 接口是对资源的具体使用，其功能是将配置信息注册到 Spring Bean 容器中。

在 Resource 资源加载器的实现中，包括 ClassPath、本地文件（File）、远程云文件（HTTP 文件），这 3 部分与 Spring 源码中的设计和实现保持一致，最终都在 DefaultResourceLoader 中调用。

BeanDefinitionReader 接口、AbstractBeanDefinitionReader 抽象类、XmlBeanDefinitionReader 实现类合理、清晰地处理了资源读取后注册 Bean 对象的操作。

本章参考 Spring 源码，设置了相应接口的继承和实现关系，如图 5-4 所示。随着框架的逐步完善，接口的作用会逐渐体现出来。

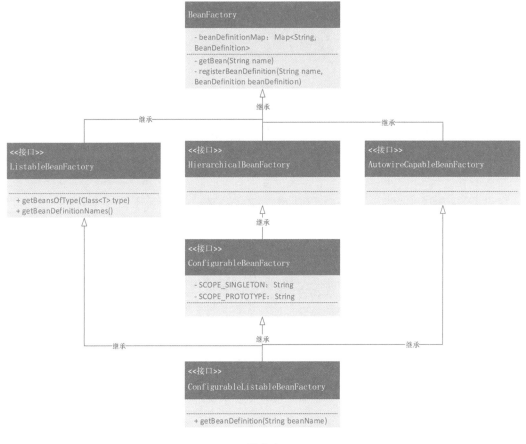

图 5-4

BeanFactory 是已经存在的 Bean 工厂接口，用于获取 Bean 对象，并新增了按照类型获取 Bean 对象的方法——<T> T getBean(String name, Class<T> requiredType)。

ListableBeanFactory 是一个接口，其功能是扩展 Bean 工厂接口，该接口中新增了 getBeansOfType 方法、getBeanDefinitionNames 方法，在 Spring 源码中还有其他扩展方法。

在 Spring 源码中，HierarchicalBeanFactory 是一个扩展 Bean 工厂层次的子接口，提供了可以获取父类 BeanFactory 的方法。

AutowireCapableBeanFactory 是一个自动化处理 Bean 工厂配置的接口，目前在实例中还没有完成相应的实现，后续会逐步完善。

ConfigurableBeanFactory 是一个可获取 BeanPostProcessor、BeanClassLoader 等方法的配置化接口。

ConfigurableListableBeanFactory 是一个提供分析和修改 Bean 对象与预先实例化的接口，不过目前只有一个 getBeanDefinition 方法。

2. 资源加载接口的定义和实现

源码详见：cn.bugstack.springframework.core.io.Resource。

```
public interface Resource {

    InputStream getInputStream() throws IOException;

}
```

在 Spring 框架下创建 core.io 核心包，主要用于处理资源加载流。

首先定义 Resource 接口，提供获取 InputStream 流的方法，然后分别实现 3 种不同的流文件——ClassPath、FileSystem 和 URL。

源码详见：cn.bugstack.springframework.core.io.ClassPathResource。

```
public class ClassPathResource implements Resource {

    private final String path;

    private ClassLoader classLoader;

    public ClassPathResource(String path) {
        this(path, (ClassLoader) null);
    }

    public ClassPathResource(String path, ClassLoader classLoader) {
        Assert.notNull(path, "Path must not be null");
        this.path = path;
        this.classLoader = (classLoader != null ? classLoader : ClassUtils.getDefaultClassLoader());
    }

    @Override
    public InputStream getInputStream() throws IOException {
        InputStream is = classLoader.getResourceAsStream(path);
```

```
        if (is == null) {
            throw new FileNotFoundException(
                    this.path + " cannot be opened because it does not exist");
        }
        return is;
    }
}
```

这部分是通过 ClassLoader 读取 ClassPath 中的文件信息实现的,具体的读取命令是 classLoader.getResourceAsStream(path)。

源码详见:cn.bugstack.springframework.core.io.FileSystemResource。

```
public class FileSystemResource implements Resource {

    private final File file;

    private final String path;

    public FileSystemResource(File file) {
        this.file = file;
        this.path = file.getPath();
    }

    public FileSystemResource(String path) {
        this.file = new File(path);
        this.path = path;
    }

    @Override
    public InputStream getInputStream() throws IOException {
        return new FileInputStream(this.file);
    }

    public final String getPath() {
        return this.path;
    }

}
```

通过指定文件路径的方式读取文件信息时会读取到一些 TXT 文件和 Excel 文件,将这些文件输出到控制台。

源码详见：cn.bugstack.springframework.core.io.UrlResource。

```java
public class UrlResource implements Resource{

    private final URL url;

    public UrlResource(URL url) {
        Assert.notNull(url,"URL must not be null");
        this.url = url;
    }

    @Override
    public InputStream getInputStream() throws IOException {
        URLConnection con = this.url.openConnection();
        try {
            return con.getInputStream();
        }
        catch (IOException ex){
            if (con instanceof HttpURLConnection){
                ((HttpURLConnection) con).disconnect();
            }
            throw ex;
        }
    }

}
```

读者可以通过 HTTP 读取远程云文件（HTTP 文件），也可以将配置文件放到 GitHub 或者 Gitee 平台中进行读取。

3. 包装资源加载器

资源加载方式有多种，资源加载器可以将这些方式放到统一的类服务下进行处理，外部用户只需要传递资源地址。

源码详见：cn.bugstack.springframework.core.io.ResourceLoader。

```java
public interface ResourceLoader {

    /**
     * Pseudo URL prefix for loading from the class path: "classpath:"
     */
    String CLASSPATH_URL_PREFIX = "classpath:";
```

```
    Resource getResource(String location);

}
```

定义获取资源的接口，在接口中传递资源地址。

源码详见：cn.bugstack.springframework.core.io.DefaultResourceLoader。

```
public class DefaultResourceLoader implements ResourceLoader {

    @Override
    public Resource getResource(String location) {
        Assert.notNull(location, "Location must not be null");
        if (location.startsWith(CLASSPATH_URL_PREFIX)) {
            return new ClassPathResource(location.substring(CLASSPATH_URL_PREFIX.length()));
        }
        else {
            try {
                URL url = new URL(location);
                return new UrlResource(url);
            } catch (MalformedURLException e) {
                return new FileSystemResource(location);
            }
        }
    }

}
```

在获取资源的过程中，主要对 3 种不同类型的资源处理方式进行了包装，分别用于判断是否为 ClassPath、FileSystem 或 URL 文件。

DefaultResourceLoader 类的实现过程比较简单，也不会让外部调用方知道过多的细节，仅关心调用结果即可。

4. Bean 对象定义读取接口

源码详见：cn.bugstack.springframework.beans.factory.support.BeanDefinitionReader。

```
public interface BeanDefinitionReader {

    BeanDefinitionRegistry getRegistry();

    ResourceLoader getResourceLoader();
```

```
    void loadBeanDefinitions(Resource resource) throws BeansException;

    void loadBeanDefinitions(Resource... resources) throws BeansException;

    void loadBeanDefinitions(String location) throws BeansException;
}
```

这是一个用于读取 Bean 对象定义的简单接口。其中定义了几个方法，如 getRegistry 方法、getResourceLoader 方法及 3 个加载 Bean 对象定义的方法。

这里需要注意 getRegistry 方法、getResourceLoader 方法，它们都为加载 Bean 对象定义的方法提供了工具，这两个方法的实现会被包装到抽象类中，以免与具体的接口实现方法产生冲突。

5. Bean 定义抽象类实现

源码详见：cn.bugstack.springframework.beans.factory.support.AbstractBeanDefinitionReader。

```java
public abstract class AbstractBeanDefinitionReader implements BeanDefinitionReader {

    private final BeanDefinitionRegistry registry;

    private ResourceLoader resourceLoader;

    protected AbstractBeanDefinitionReader(BeanDefinitionRegistry registry) {
        this(registry, new DefaultResourceLoader());
    }

    public AbstractBeanDefinitionReader(BeanDefinitionRegistry registry, ResourceLoader resourceLoader) {
        this.registry = registry;
        this.resourceLoader = resourceLoader;
    }

    @Override
    public BeanDefinitionRegistry getRegistry() {
        return registry;
    }

    @Override
    public ResourceLoader getResourceLoader() {
        return resourceLoader;
```

    }

}

抽象类实现了 BeanDefinitionReader 接口的前两个方法，并提供了构造函数，此时可以通过外部调用将 Bean 对象的定义注入并传递到类中。

这样在 BeanDefinitionReader 接口的具体实现类中，就可以将解析后的 XML 文件中的 Bean 对象信息注册到 Spring Bean 容器中。在之前的单元测试中，我们通过调用 BeanDefinitionRegistry 完成了 Bean 对象的注册，现在可以放到 XML 文件中了。

6. 解析 XML 处理 Bean 注册

源码详见：cn.bugstack.springframework.beans.factory.xml.XmlBeanDefinitionReader。

```java
public class XmlBeanDefinitionReader extends AbstractBeanDefinitionReader {

    public XmlBeanDefinitionReader(BeanDefinitionRegistry registry) {
        super(registry);
    }

    public XmlBeanDefinitionReader(BeanDefinitionRegistry registry, ResourceLoader resourceLoader) {
        super(registry, resourceLoader);
    }

    @Override
    public void loadBeanDefinitions(Resource resource) throws BeansException {
        try {
            try (InputStream inputStream = resource.getInputStream()) {
                doLoadBeanDefinitions(inputStream);
            }
        } catch (IOException | ClassNotFoundException e) {
            throw new BeansException("IOException parsing XML document from " + resource, e);
        }
    }

    @Override
    public void loadBeanDefinitions(Resource... resources) throws BeansException {
        for (Resource resource : resources) {
            loadBeanDefinitions(resource);
        }
    }
```

```java
@Override
public void loadBeanDefinitions(String location) throws BeansException {
    ResourceLoader resourceLoader = getResourceLoader();
    Resource resource = resourceLoader.getResource(location);
    loadBeanDefinitions(resource);
}

protected void doLoadBeanDefinitions(InputStream inputStream) throws ClassNotFoundException {
    Document doc = XmlUtil.readXML(inputStream);
    Element root = doc.getDocumentElement();
    NodeList childNodes = root.getChildNodes();

    for (int i = 0; i < childNodes.getLength(); i++) {
        // 判断元素
        if (!(childNodes.item(i) instanceof Element)) continue;
        // 判断对象
        if (!"bean".equals(childNodes.item(i).getNodeName())) continue;

        // 解析标签
        Element bean = (Element) childNodes.item(i);
        String id = bean.getAttribute("id");
        String name = bean.getAttribute("name");
        String className = bean.getAttribute("class");
        // 获取 Class，方便获取类中的名称
        Class<?> clazz = Class.forName(className);
        // 优先级 id>name
        String beanName = StrUtil.isNotEmpty(id) ? id : name;
        if (StrUtil.isEmpty(beanName)) {
            beanName = StrUtil.lowerFirst(clazz.getSimpleName());
        }

        // 定义 Bean 对象
        BeanDefinition beanDefinition = new BeanDefinition(clazz);
        // 读取属性并填充
        for (int j = 0; j < bean.getChildNodes().getLength(); j++) {
            if (!(bean.getChildNodes().item(j) instanceof Element)) continue;
            if (!"property".equals(bean.getChildNodes().item(j).getNodeName()))
            continue;
            // 解析标签：property
            Element property = (Element) bean.getChildNodes().item(j);
            String attrName = property.getAttribute("name");
```

```
                String attrValue = property.getAttribute("value");
                String attrRef = property.getAttribute("ref");
                // 获取属性值：引入对象、值对象
                Object value = StrUtil.isNotEmpty(attrRef) ? new BeanReference(attrRef):
                attrValue;
                // 创建属性信息
                PropertyValue propertyValue = new PropertyValue(attrName, value);
                beanDefinition.getPropertyValues().addPropertyValue(propertyValue);
            }
            if (getRegistry().containsBeanDefinition(beanName)) {
                throw new BeansException("Duplicate beanName[" + beanName + "] is not
allowed");
            }
            // 注册 BeanDefinition
            getRegistry().registerBeanDefinition(beanName, beanDefinition);
        }
    }
}
```

XmlBeanDefinitionReader 类最核心的内容就是对 XML 文件的解析，通过解析 XML 文件自动注册的方式来实现。

loadBeanDefinitions 方法用于处理资源加载。这里新增了一个内部方法 doLoadBeanDefinitions，其主要功能是解析 XML 文件。

doLoadBeanDefinitions 方法主要是对 XML 文件进行读取和对 Element 元素进行解析。在解析过程中，我们循环获取 Bean 对象的配置，以及配置中的 id、name、class、value 和 ref 信息。

先将读取出来的配置信息创建成 BeanDefinition 及 PropertyValues，再将完整的 Bean 定义内容注册到 Spring Bean 容器 getRegistry().registerBeanDefinition(beanName, beanDefinition) 中。

5.4 配置 Bean 对象注册测试

1. 事先准备

源码详见：cn.bugstack.springframework.test.bean.UserDao。

```java
public class UserDao {

    private static Map<String, String> hashMap = new HashMap<>();

    static {
        hashMap.put("10001", "小傅哥");
        hashMap.put("10002", "八杯水");
        hashMap.put("10003", "阿毛");
    }

    public String queryUserName(String uId) {
        return hashMap.get(uId);
    }

}
```

源码详见：cn.bugstack.springframework.test.bean.UserService。

```java
public class UserService {

    private String uId;

    private UserDao userDao;

    public void queryUserInfo() {
        return userDao.queryUserName(uId);
    }

    // ...get/set
}
```

Dao、Service 是我们开发时经常会使用的类，在 UserService 中注入 UserDao，就能体现出 Bean 属性的依赖。

2. 配置文件

配置详见：important.properties。

```
# Config File
system.key=OLpj9823dZ
```

配置详见：spring.xml。

```xml
<?xml version="1.0" encoding="UTF-8"?>
<beans>
```

```xml
    <bean id="userDao" class="cn.bugstack.springframework.test.bean.UserDao"/>

    <bean id="userService" class="cn.bugstack.springframework.test.bean.UserService">
        <property name="uId" value="10001"/>
        <property name="userDao" ref="userDao"/>
    </bean>

</beans>
```

这里有两个配置文件，一个用于测试资源加载器，另一个用于测试整体的 Bean 对象的注册功能。

3. 单元测试（资源加载）

（1）测试实例。

```java
private DefaultResourceLoader resourceLoader;

@Before
public void init() {
    resourceLoader = new DefaultResourceLoader();
}

@Test
public void test_classpath() throws IOException {
    Resource resource = resourceLoader.getResource("classpath:important.properties");
    InputStream inputStream = resource.getInputStream();
    String content = IoUtil.readUtf8(inputStream);
    System.out.println(content);
}

@Test
public void test_file() throws IOException {
    Resource resource = resourceLoader.getResource("src/test/resources/important.properties");
    InputStream inputStream = resource.getInputStream();
    String content = IoUtil.readUtf8(inputStream);
    System.out.println(content);
}

@Test
public void test_url() throws IOException {
    Resource resource = resourceLoader.getResource("https://github.com/fuzhengwei/small-spring/important.properties"
```

```
        InputStream inputStream = resource.getInputStream();
        String content = IoUtil.readUtf8(inputStream);
        System.out.println(content);
}
```

（2）测试结果。

```
# Config File
system.key=OLpj9823dZ

Process finished with exit code 0
```

test_classpath、test_file、test_url 这 3 个方法分别用于测试加载 ClassPath 文件、FileSystem 文件和 URL 文件。

4．单元测试（读取配置文件并注册 Bean 对象）

（1）测试实例。

```
@Test
public void test_xml() {
    // 1. 初始化 BeanFactory 接口
    DefaultListableBeanFactory beanFactory = new DefaultListableBeanFactory();

    // 2. 读取配置文件 & 注册 Bean 对象
    XmlBeanDefinitionReader reader = new XmlBeanDefinitionReader(beanFactory);
    reader.loadBeanDefinitions("classpath:spring.xml");

    // 3. 获取 Bean 对象的调用方法
    UserService userService = beanFactory.getBean("userService", UserService.class);
    String result = userService.queryUserInfo();
    System.out.println(" 测试结果: " + result);
}
```

（2）测试结果。

```
测试结果：小傅哥

Process finished with exit code 0
```

在上面的测试实例中可以看到，将之前通过手动注册 Bean 对象及配置属性信息的内容，修改为通过 new XmlBeanDefinitionReader(beanFactory) 类读取 spring.xml 配置文件的方式来实现，并通过了测试验证。

5.5 本章总结

此时 Spring 框架的工程结构已经越来越完整了，先在配置文件中解析并注册 Bean 的信息，再通过 Bean 工厂获取 Bean 对象，并进行相应的调用。

关于实例中实现的每一个步骤，都会尽可能参照 Spring 源码中的接口定义、抽象类实现、名称规范和代码结构等知识进行相应的简化处理。读者也可以通过类名、接口或者整体的结构来学习 Spring 源码，便于掌握。

第 6 章 实现应用上下文

随着对 Spring 框架的深入学习，我们已经开始接触核心的流程设计和处理部分，其难度与我们把控一个具有复杂需求的研发设计和实现过程是类似的。解决这类复杂场景的设计主要分为——分治、抽象和知识 3 个方面，运用架构和设计模式的知识，在分治层面将一个大问题分为若干个子问题，而问题越小就越容易被理解和处理。

本章将在 Spring 框架中继续扩展新的功能。例如，在对一个 Bean 对象进行定义和实例化的过程中，是否可以满足自定义扩展需求，此时需要引入什么样的界限上下文，对 Bean 对象执行修改、记录和替换等动作呢？

- 本章难度：★★★★☆
- 本章重点：引入应用上下文，进行资源扫描与加载，为 Bean 对象实例化过程添加扩展机制，允许加载 Bean 对象和在其实例化前后进行修改和扩展。

6.1 分治 Bean 对象功能

如果在工作中开发过基于 Spring 的技术组件，或者学习过关于 SpringBoot 中间件的设计和开发等内容，那么一定做过以下内容：继承或者实现了 Spring 对外暴露的类或接口，在接口的实现中获取 BeanFactory 及 Bean 对象等内容，如修改 Bean 的信息、添加日志打印、处理数据库路由对数据源进行切换、给 RPC 服务连接注册中心等。

在对 Bean 对象进行实例化的过程中，不仅需要添加扩展机制，还需要优化 spring.

xml 配置文件的初始化和加载策略（因为不能将面向 Spring 本身开发的 DefaultListable BeanFactory 服务直接给用户使用）。具体的修改如图 6-1 所示。

```
@Test
public void test_BeanFactoryPostProcessorAndBeanPostProcessor(){
    // 1.初始化 BeanFactory接口
    DefaultListableBeanFactory beanFactory = new DefaultListableBeanFactory();

    // 2. 读取配置文件&注册Bean对象
    XmlBeanDefinitionReader reader = new XmlBeanDefinitionReader(beanFactory);
    reader.loadBeanDefinitions( location: "classpath:spring.xml");

    // 3. 在BeanDefinition 加载完成 & Bean对象实例化之前，修改 BeanDefinition 的属性值
    MyBeanFactoryPostProcessor beanFactoryPostProcessor = new MyBeanFactoryPostProcessor();
    beanFactoryPostProcessor.postProcessBeanFactory(beanFactory);
```
步骤合并

图 6-1

在目前的 Spring 框架中，DefaultListableBeanFactory、XmlBeanDefinitionReader 是我们对于服务功能进行测试时的使用方法。它们能很好地体现出 Spring 框架是如何加载 XML 文件及注册 Bean 对象的，但这种方法是面向 Spring 框架本身的，不具备一定的扩展性。

就像需要在 Bean 初始化过程中完成对 Bean 对象的扩展，但是很难做到自动化处理。所以我们要将 Bean 对象的扩展机制功能和对 Spring 框架上下文的包装进行整合，以提供完整的服务。

6.2　Bean 对象扩展和上下文设计

为了在 Bean 对象从注册到实例化的过程中执行用户的自定义操作，需要在 Bean 的定义和初始化过程中插入接口类，这个接口类再由外部实现自己需要的服务事项。再结合 Spring 框架上下文的应用，就可以满足我们的目标需求。整体设计结构如图 6-2 所示。

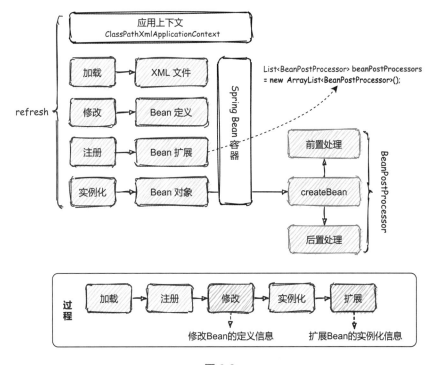

图 6-2

满足 Bean 对象扩展的 BeanFactoryPostProcessor 和 BeanPostProcessor 是 Spring 框架非常重量级的两个接口，也是使用 Spring 框架新增开发组件需求的两个必备接口。

BeanFactoryPostProcessor 接口是由 Spring 框架组件提供的容器扩展机制，允许在 Bean 对象注册后、未实例化之前，修改 Bean 对象的定义信息 BeanDefinition。

BeanPostProcessor 接口也是 Spring 提供的扩展机制，不同的是 BeanPostProcessor 接口是在 Bean 对象执行初始化方法前后，对 Bean 对象进行修改、记录、替换等。这部分扩展功能与后面 AOP 的实现有着密切的关系。

如果只添加 BeanFactoryPostProcessor 和 BeanPostProcessor 这两个接口，不做任何包装，那么使用时是非常困难的。我们的目的是开发 Spring 的上下文类，将相应的 XML 文件的加载、注册、实例化及新增的修改和扩展功能全部融合进去，使 Spring 可以自动扫描到新增的服务，便于用户使用。

6.3 Bean 对象扩展和上下文实现

1. 工程结构

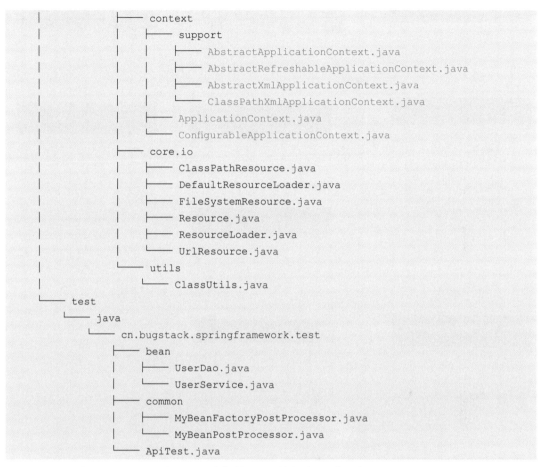

Spring 应用上下文和对 Bean 对象扩展机制的类的关系如图 6-3 所示。

整个类图主要体现的是关于 Spring 应用上下文及对 Bean 对象扩展机制的实现。

从继承了 ListableBeanFactory 接口的 ApplicationContext 接口开始，扩展出一系列应用上下文的抽象实现类，并最终完成 ClassPathXmlApplicationContext 类的实现。这个类就是最后用户使用的类。

同时，在实现应用上下文的过程中，通过定义 BeanFactoryPostProcessor、BeanPostProcessor 两个接口，实现对 Bean 对象的扩展机制。

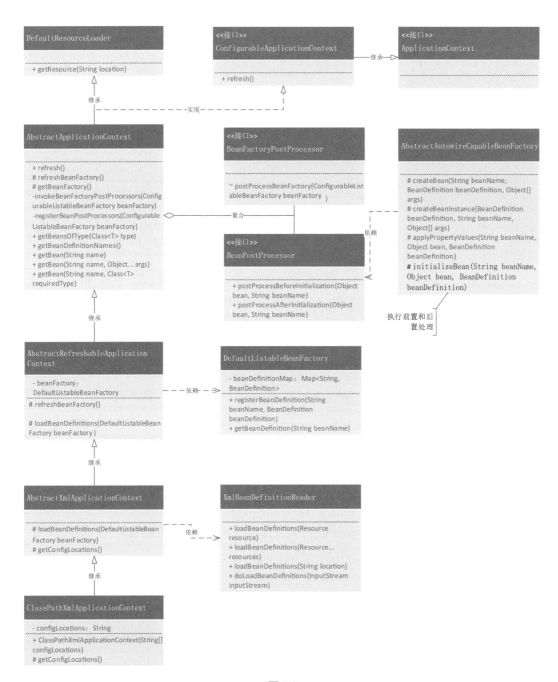

图 6-3

2. 定义 BeanFactoryPostProcessor 接口

源码详见：cn.bugstack.springframework.beans.factory.config.BeanFactoryPostProcessor。

```java
public interface BeanFactoryPostProcessor {

    /**
     * 在所有的 BeanDefinition 加载完成后，且将 Bean 对象实例化之前，提供修改 BeanDefinition
     * 属性的机制
     *
     * @param beanFactory
     * @throws BeansException
     */
    void postProcessBeanFactory(ConfigurableListableBeanFactory beanFactory) throws BeansException;

}
```

在 Spring 源码中有这样一段描述："Allows for custom modification of an application context's bean definitions,adapting the bean property values of the context's underlying bean factory"。即 BeanFactoryPostProcessor 接口是在所有的 BeanDefinition 加载完成后，且将 Bean 对象实例化之前，提供修改 BeanDefinition 属性的机制。

3. 定义 BeanPostProcessor 接口

源码详见：cn.bugstack.springframework.beans.factory.config.BeanPostProcessor。

```java
public interface BeanPostProcessor {

    /**
     * 在 Bean 对象执行初始化方法之前，执行此方法
     *
     * @param bean
     * @param beanName
     * @return
     * @throws BeansException
     */
    Object postProcessBeforeInitialization(Object bean, String beanName) throws BeansException;

    /**
     * 在 Bean 对象执行初始化方法之后，执行此方法
     *
     * @param bean
     * @param beanName
```

```
     * @return
     * @throws BeansException
     */
    Object postProcessAfterInitialization(Object bean, String beanName) throws BeansException;
}
```

在 Spring 源码中有这样一段描述:"Factory hook that allows for custom modification of new bean instances,e.g. checking for marker interfaces or wrapping them with proxies"。即 BeanPostProcessor 接口提供了修改新实例化 Bean 对象的扩展点。

另外此接口提供了两个方法,它们的作用分别是:postProcessBeforeInitialization 方法用于在 Bean 对象执行初始化方法之前,执行此方法;postProcessAfterInitialization 方法用于在 Bean 对象执行初始化方法之后,执行此方法。

4. 定义上下文接口

源码详见:cn.bugstack.springframework.context.ApplicationContext。

```
public interface ApplicationContext extends ListableBeanFactory {
}
```

context 是为本次实现应用上下文功能而新增的服务包。

ApplicationContext 继承于 ListableBeanFactory,即继承了关于 BeanFactory 方法,如 getBean 的一些方法。由于 ApplicationContext 本身是接口,但目前不需要添加获取 ID 和父类上下文的方法,所以暂时没有该接口方法的定义。

源码详见:cn.bugstack.springframework.context.ConfigurableApplicationContext。

```
public interface ConfigurableApplicationContext extends ApplicationContext {

    /**
     * 刷新容器
     *
     * @throws BeansException
     */
    void refresh() throws BeansException;

}
```

ConfigurableApplicationContext 继承于 ApplicationContext,并提供了核心方法 refresh。接下来也需要在上下文的实现过程中刷新容器。

5. 应用上下文抽象类实现

源码详见：cn.bugstack.springframework.context.support.AbstractApplicationContext。

```java
public abstract class AbstractApplicationContext extends DefaultResourceLoader
implements ConfigurableApplicationContext {

    @Override
    public void refresh() throws BeansException {
        // 1. 创建 BeanFactory，并加载 BeanDefinition
        refreshBeanFactory();

        // 2. 获取 BeanFactory
        ConfigurableListableBeanFactory beanFactory = getBeanFactory();

        // 3. 在将 Bean 对象实例化之前，执行 BeanFactoryPostProcessor 操作
        invokeBeanFactoryPostProcessors(beanFactory);

        // 4. BeanPostProcessor 需要在将 Bean 对象实例化之前注册
        registerBeanPostProcessors(beanFactory);

        // 5. 提前实例化单例 Bean 对象
        beanFactory.preInstantiateSingletons();
    }

    protected abstract void refreshBeanFactory() throws BeansException;

    protected abstract ConfigurableListableBeanFactory getBeanFactory();

    private void invokeBeanFactoryPostProcessors(ConfigurableListableBeanFactory beanFactory) {
        Map<String, BeanFactoryPostProcessor> beanFactoryPostProcessorMap = beanFactory.getBeansOfType(BeanFactoryPostProcessor.class);
        for (BeanFactoryPostProcessor beanFactoryPostProcessor: beanFactoryPostProcessorMap.values()) {
            beanFactoryPostProcessor.postProcessBeanFactory(beanFactory);
        }
    }

    private void registerBeanPostProcessors(ConfigurableListableBeanFactory beanFactory) {
        Map<String, BeanPostProcessor> beanPostProcessorMap = beanFactory.getBeansOfType(BeanPostProcessor.class);
        for (BeanPostProcessor beanPostProcessor : beanPostProcessorMap.values()) {
            beanFactory.addBeanPostProcessor(beanPostProcessor);
        }
```

```
    }

    //... getBean、getBeansOfType、getBeanDefinitionNames 方法

}
```

AbstractApplicationContext 继承 DefaultResourceLoader 是为了处理 spring.xml 配置文件中配置资源的加载。

refresh 方法的定义及实现过程如下。

- 创建 BeanFactory，并加载 BeanDefinition。
- 获取 BeanFactory。
- 在将 Bean 对象实例化之前，执行 BeanFactoryPostProcessor 操作。
- BeanPostProcessor 需要在将 Bean 对象实例化之前注册。
- 提前实例化单例 Bean 对象。
- 将定义的抽象方法 refreshBeanFactory 和 getBeanFactory 由继承此抽象类的其他抽象类来实现。

6. 获取 Bean 工厂和加载资源

源码详见：cn.bugstack.springframework.context.support.AbstractRefreshableApplicationContext。

```
public abstract class AbstractRefreshableApplicationContext extends AbstractApplicationContext {

    private DefaultListableBeanFactory beanFactory;

    @Override
    protected void refreshBeanFactory() throws BeansException {
        DefaultListableBeanFactory beanFactory = createBeanFactory();
        loadBeanDefinitions(beanFactory);
        this.beanFactory = beanFactory;
    }

    private DefaultListableBeanFactory createBeanFactory() {
        return new DefaultListableBeanFactory();
    }
```

```
    protected abstract void loadBeanDefinitions(DefaultListableBeanFactory beanFactory);

    @Override
    protected ConfigurableListableBeanFactory getBeanFactory() {
        return beanFactory;
    }

}
```

使用 refreshBeanFactory 抽象方法可以获取 DefaultListableBeanFactory 的实例化，以及对资源配置的加载 loadBeanDefinitions(beanFactory)，在加载完成后即可完成对 spring.xml 配置文件中 Bean 对象的定义和注册，也实现了 BeanFactoryPostProcessor 接口、BeanPostProcessor 接口对 Bean 信息的配置。

此时的资源加载只定义了一个抽象类方法 loadBeanDefinitions(DefaultListableBeanFactory beanFactory)，其余步骤由其他抽象类继承实现。

7. 上下文中对配置信息的加载

源码详见：cn.bugstack.springframework.context.support.AbstractXmlApplicationContext。

```
public abstract class AbstractXmlApplicationContext extends AbstractRefreshableApplicationContext {

    @Override
    protected void loadBeanDefinitions(DefaultListableBeanFactory beanFactory) {
        XmlBeanDefinitionReader beanDefinitionReader = new XmlBeanDefinitionReader(beanFactory, this);
        String[] configLocations = getConfigLocations();
        if (null != configLocations){
            beanDefinitionReader.loadBeanDefinitions(configLocations);
        }
    }

    protected abstract String[] getConfigLocations();

}
```

在 AbstractXmlApplicationContext 抽象类的 loadBeanDefinitions 方法中，使用 XmlBeanDefinitionReader 类来处理 XML 文件中的配置信息。

同时这里又有一个抽象类方法 getConfigLocations，此方法是为了从入口上下文的类中取得配置信息的地址。

8. 应用上下文实现类（ClassPathXmlApplicationContext）

源码详见：cn.bugstack.springframework.context.support.ClassPathXmlApplicationContext。

```java
public class ClassPathXmlApplicationContext extends AbstractXmlApplicationContext {

    private String[] configLocations;

    public ClassPathXmlApplicationContext() {
    }

    /**
     * 从 XML 文件中加载 BeanDefinition，并刷新上下文
     *
     * @param configLocations
     * @throws BeansException
     */
    public ClassPathXmlApplicationContext(String configLocations) throws BeansException {
        this(new String[]{configLocations});
    }

    /**
     * 从 XML 文件中加载 BeanDefinition，并刷新上下文
     * @param configLocations
     * @throws BeansException
     */
    public ClassPathXmlApplicationContext(String[] configLocations) throws BeansException {
        this.configLocations = configLocations;
        refresh();
    }

    @Override
    protected String[] getConfigLocations() {
        return configLocations;
    }

}
```

ClassPathXmlApplicationContext 是具体对外给用户提供的应用上下文类。

在继承了 AbstractXmlApplicationContext 类及多层抽象类的功能后，ClassPathXmlApplicationContext 类的实现就相对简单了，主要调用了继承抽象类中的方法和提供了配置文件的地址信息。

9. 在创建 Bean 对象时完成前置和后置处理

源码详见：cn.bugstack.springframework.beans.factory.support.AbstractAutowireCapableBeanFactory。

```java
public abstract class AbstractAutowireCapableBeanFactory extends AbstractBeanFactory
        implements AutowireCapableBeanFactory {

    private InstantiationStrategy instantiationStrategy = new CglibSubclassingInstantiationStrategy();

    @Override
    protected Object createBean(String beanName, BeanDefinition beanDefinition, Object[] args) throws BeansException {
        Object bean = null;
        try {
            bean = createBeanInstance(beanDefinition, beanName, args);
            // 给 Bean 对象填充属性
            applyPropertyValues(beanName, bean, beanDefinition);
            // 执行 Bean 对象的初始化方法和 BeanPostProcessor 接口的前置和后置处理方法
            bean = initializeBean(beanName, bean, beanDefinition);
        } catch (Exception e) {
            throw new BeansException("Instantiation of bean failed", e);
        }

        registerSingleton(beanName, bean);
        return bean;
    }

    public InstantiationStrategy getInstantiationStrategy() {
        return instantiationStrategy;
    }

    public void setInstantiationStrategy(InstantiationStrategy instantiationStrategy) {
        this.instantiationStrategy = instantiationStrategy;
    }

    private Object initializeBean(String beanName, Object bean, BeanDefinition beanDefinition) {
        // 1. 执行 BeanPostProcessor Before 前置处理
        Object wrappedBean = applyBeanPostProcessorsBeforeInitialization(bean, beanName);

        // 待完成内容
        invokeInitMethods(beanName, wrappedBean, beanDefinition);
```

```java
        // 2. 执行 BeanPostProcessor After 后置处理
        wrappedBean = applyBeanPostProcessorsAfterInitialization(bean, beanName);
        return wrappedBean;
    }

    private void invokeInitMethods(String beanName, Object wrappedBean, BeanDefinition beanDefinition) {

    }

    @Override
    public Object applyBeanPostProcessorsBeforeInitialization(Object existingBean, String beanName) throws BeansException {
        Object result = existingBean;
        for (BeanPostProcessor processor : getBeanPostProcessors()) {
            Object current = processor.postProcessBeforeInitialization(result, beanName);
            if (null == current) return result;
            result = current;
        }
        return result;
    }

    @Override
    public Object applyBeanPostProcessorsAfterInitialization(Object existingBean, String beanName) throws BeansException {
        Object result = existingBean;
        for (BeanPostProcessor processor : getBeanPostProcessors()) {
            Object current = processor.postProcessAfterInitialization(result, beanName);
            if (null == current) return result;
            result = current;
        }
        return result;
    }
}
```

当实现BeanPostProcessor接口后，会涉及两个接口方法——postProcessBeforeInitialization、postProcessAfterInitialization，分别用于进行 Bean 对象执行初始化前后的处理。

也就是说在创建 Bean 对象时，在 createBean 方法中添加了 initializeBean(beanName, bean, beanDefinition)，而这个操作主要是使用 applyBeanPostProcessorsBeforeInitialization 方法

和 applyBeanPostProcessorsAfterInitialization 方法实现的。

此外，applyBeanPostProcessorsBeforeInitialization 方法和 applyBeanPostProcessorsAfterInitialization 方法是在 AutowireCapableBeanFactory 接口类中新增加的两个方法。

6.4 应用上下文功能测试

1. 事先准备

源码详见：cn.bugstack.springframework.test.bean.UserDao。

```
public class UserDao {

    private static Map<String, String> hashMap = new HashMap<>();

    static {
        hashMap.put("10001", " 小傅哥 ");
        hashMap.put("10002", " 八杯水 ");
        hashMap.put("10003", " 阿毛 ");
    }

    public String queryUserName(String uId) {
        return hashMap.get(uId);
    }

}
```

源码详见：cn.bugstack.springframework.test.bean.UserService。

```
public class UserService {

    private String uId;
    private String company;
    private String location;
    private UserDao userDao;

    public void queryUserInfo() {
        return userDao.queryUserName(uId);
    }

    // ...get/set
}
```

这里新增了 company 和 location 两个属性信息，用于测试 BeanPostProcessor 和 BeanFactoryPostProcessor 两个接口对 Bean 属性信息的扩展功能。

2. 实现 BeanPostProcessor 类和 BeanFactoryPostProcessor 类

源码详见：cn.bugstack.springframework.test.common.MyBeanFactoryPostProcessor。

```java
public class MyBeanFactoryPostProcessor implements BeanFactoryPostProcessor {

    @Override
    public void postProcessBeanFactory(ConfigurableListableBeanFactory beanFactory) throws BeansException {

        BeanDefinition beanDefinition = beanFactory.getBeanDefinition("userService");
        PropertyValues propertyValues = beanDefinition.getPropertyValues();

        propertyValues.addPropertyValue(new PropertyValue("company", "改为：字节跳动"));
    }

}
```

源码详见：cn.bugstack.springframework.test.common.MyBeanPostProcessor。

```java
public class MyBeanPostProcessor implements BeanPostProcessor {

    @Override
    public Object postProcessBeforeInitialization(Object bean, String beanName) throws BeansException {
        if ("userService".equals(beanName)) {
            UserService userService = (UserService) bean;
            userService.setLocation("改为：北京");
        }
        return bean;
    }

    @Override
    public Object postProcessAfterInitialization(Object bean, String beanName) throws BeansException {
        return bean;
    }

}
```

如果在 Spring 中进行了一些组件的开发，那么一定非常熟悉 BeanPostProcessor 和

BeanFactoryPostProcessor 这两个类。该测试主要是实现这两个类，并对实例化过程中的 Bean 对象完成某些操作。

3. 配置文件

基础配置文件。不包含 BeanFactoryPostProcessor 和 BeanPostProcessor 实现类。

```xml
<?xml version="1.0" encoding="UTF-8"?>
<beans>

    <bean id="userDao" class="cn.bugstack.springframework.test.bean.UserDao"/>

    <bean id="userService" class="cn.bugstack.springframework.test.bean.UserService">
        <property name="uId" value="10001"/>
        <property name="company" value="腾讯"/>
        <property name="location" value="深圳"/>
        <property name="userDao" ref="userDao"/>
    </bean>

</beans>
```

增强配置文件。包含 BeanFactoryPostProcessor 和 BeanPostProcessor 实现类。

```xml
<?xml version="1.0" encoding="UTF-8"?>
<beans>

    <bean id="userDao" class="cn.bugstack.springframework.test.bean.UserDao"/>

    <bean id="userService" class="cn.bugstack.springframework.test.bean.UserService">
        <property name="uId" value="10001"/>
        <property name="company" value="腾讯"/>
        <property name="location" value="深圳"/>
        <property name="userDao" ref="userDao"/>
    </bean>

    <bean class="cn.bugstack.springframework.test.common.MyBeanPostProcessor"/>
    <bean class="cn.bugstack.springframework.test.common.MyBeanFactoryPostProcessor"/>

</beans>
```

这里提供了两个配置文件，一个是不包含 BeanFactoryPostProcessor 和 BeanPostProcessor 实现类的基础配置文件，另一个是包含两者的增强配置文件。之所以这样配置，主要是对比验证在运用或不运用 Spring 新增加的应用上下文时，它们都是怎么实现的。

4. 不使用应用上下文

```
@Test
public void test_BeanFactoryPostProcessorAndBeanPostProcessor(){
    // 1. 初始化 BeanFactory 接口
    DefaultListableBeanFactory beanFactory = new DefaultListableBeanFactory();

    // 2. 读取配置文件，注册 Bean 对象
    XmlBeanDefinitionReader reader = new XmlBeanDefinitionReader(beanFactory);
    reader.loadBeanDefinitions("classpath:spring.xml");

    // 3. BeanDefinition 加载完成，在将 Bean 对象实例化之前，修改 BeanDefinition 的属性值
    MyBeanFactoryPostProcessor beanFactoryPostProcessor = new MyBeanFactoryPostProcessor();
    beanFactoryPostProcessor.postProcessBeanFactory(beanFactory);

    // 4. 在 Bean 对象实例化之后，修改 Bean 对象的属性信息
    MyBeanPostProcessor beanPostProcessor = new MyBeanPostProcessor();
    beanFactory.addBeanPostProcessor(beanPostProcessor);

    // 5. 获取 Bean 对象的调用方法
    UserService userService = beanFactory.getBean("userService", UserService.class);
    String result = userService.queryUserInfo();
    System.out.println("测试结果:" + result);
}
```

首先，使用 DefaultListableBeanFactory 类创建 BeanFactory 接口并使用 XmlBeanDefinition Reader 类加载配置文件。

然后，对 MyBeanFactoryPostProcessor 类和 MyBeanPostProcessor 类进行处理，前者是在 BeanDefinition 加载完成且在将 Bean 对象实例化之前修改 BeanDefinition 的属性值，后者是在将 Bean 对象实例化之后修改 Bean 对象的属性信息。

测试结果如下。

```
测试结果：小傅哥，改为：字节跳动，改为：北京

Process finished with exit code 0
```

从测试结果中可以看到，配置的属性信息已经与 spring.xml 配置文件中不一样了。

5. 使用应用上下文

```
@Test
public void test_xml() {
    // 1. 初始化 BeanFactory 接口
```

```
    ClassPathXmlApplicationContext applicationContext = new ClassPathXmlApplicationContext
("classpath:springPostProcessor.xml");

    // 2. 获取 Bean 对象的调用方法
    UserService userService = applicationContext.getBean("userService", UserService.class);
    String result = userService.queryUserInfo();
    System.out.println("测试结果: " + result);
}
```

使用新增的 ClassPathXmlApplicationContext 应用上下文类会更加方便。在这里可以将配置文件这一步骤交给 ClassPathXmlApplicationContext 类完成，也不需要管理一些通过自定义实现的 Spring 接口的类。

测试结果如下。

```
测试结果：小傅哥，改为：字节跳动，改为：北京

Process finished with exit code 0
```

从测试结果中可以看到，此次的测试结果与不使用应用上下文的测试结果是一致的，但是，使用该方法会更加简单。

6.5 本章总结

本章新增了 Spring 框架中两个非常重要的接口——BeanFactoryPostProcessor、BeanPostProcessor，还添加了关于应用上下文的实现。ApplicationContext 接口的定义是继承 BeanFactory 新增加功能的接口，具有自动识别、资源加载、容器事件和事件监听器等功能。同时，对于一些国际化支持、单例 Bean 自动初始化等，也可以在 ApplicationContext 接口中实现和扩充。

读者通过本章的讲解，应该已经了解了 BeanFactoryPostProcessor 接口和 BeanPostProcessor 接口，以后再开发 Spring 中间件时，如果需要获取 Bean 对象及修改一些属性信息，就可以使用这两个接口。同时，BeanPostProcessor 接口也是实现 AOP 切面技术的关键。

第 7 章
Bean 对象的初始化和销毁

在逐步完善 Spring 框架的过程中，相信大家已经了解了面向对象开发的特性，如封装、继承、多态等。这些特性通过定义接口、接口继承接口、由抽象类实现接口、类继承的类实现了接口方法，使程序逻辑做到分层、分区和分块，并将核心逻辑层和功能的使用进行封装隔离。当功能需要变更迭代时，只需要在合适的层完成装配即可，不会影响核心逻辑。

接下来继续完善 Bean 在 Spring Bean 容器中的生命周期，使 Bean 对象既可以随着程序的启动在 Spring Bean 容器中进行初始化，也可以在退出程序时，对 Bean 对象进行销毁。这时会使用 JVM 的注册钩子功能，使虚拟机在关闭之前销毁。

- 本章难度：★★★☆☆
- 本章重点：在 XML 配置中添加注解 init-method、destroy-method，扩展 Bean 对象在实例化过程中的初始化，以及向虚拟机注册钩子并在退出程序时销毁。

7.1 容器管理 Bean 功能

当我们将使用类创建的 Bean 对象交给 Spring Bean 容器管理后，这个对象就可以被赋予更多的使用功能。就像在前文的实例开发中，已经为 Bean 对象的实例化过程，扩展了在 Bean 对象未被实例化前的属性信息修改及初始化完成前后的对象处理。对 Bean 对象的扩展功能，都可以让我们随系统开发诉求对工程中的 Bean 对象做出相应的改变。

我们还可以在 Bean 对象的初始化过程中执行一些其他操作，如加载数据，连接注册中心暴露 RPC 接口，以及在 Web 程序关闭时执行断开链接、销毁内存等。如果没有 Spring Bean 容器，则可以通过构造函数、静态方法及手动调用的方式实现，但这种处理方式不如将操作都交给 Spring Bean 容器管理更合适。此时，我们会看到 spring.xml 配置文件中有如下操作，如图 7-1 所示。

```
<beans>
    <bean id="userDao" class="cn.bugstack.springframework.test.bean.UserDao"
        init-method="initDataMethod"          初始化方法
        destroy-method="destroyDataMethod"/>  销毁方法

    <bean id="userService" class="cn.bugstack.springframework.test.bean.UserService">
        <property name="uId" value="10001"/>
        <property name="company" value="腾讯"/>
        <property name="location" value="深圳"/>
        <property name="userDao" ref="userDao"/>
    </bean>
</beans>
```

图 7-1

对于满足用户在 XML 中配置初始化和销毁的方法，也可以通过实现类的方式自行包装处理，如在使用 Spring Bean 容器时用到的 InitializingBean 和 DisposableBean 两个接口。

还有一种采用注解的方法进行初始化，不过目前还没有涉及注解的知识，后续再完善此类功能。

7.2 初始化和销毁设计

将对外暴露的接口进行定义使用或者 XML 配置，完成一系列的扩展，都会让 Spring 框架看上去很神秘。对于在 Spring Bean 容器初始化过程中额外添加的处理有两种方法：第一种是通过依赖倒置的方式预先执行了一个定义好的接口方法；第二种是反射调用类中 XML 配置的方法，最终只要按照接口定义实现，就会由 Spring Bean 容器在处理 Bean

对象注册的过程中进行调用。整体设计结构如图 7-2 所示。

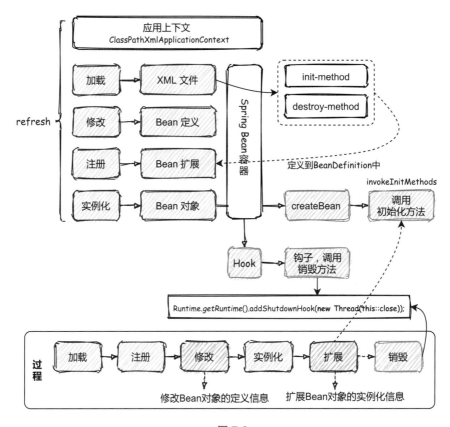

图 7-2

首先，在 spring.xml 配置文件中添加 init-method 和 destroy-method 两个注解；然后，在加载配置文件的过程中，将注解一并定义到 BeanDefinition 的属性中。这样在 initializeBean 初始化过程中，就可以通过反射的方式调用在 BeanDefinition 属性中配置的初始化和销毁方法。如果采用的是接口实现的方式，则可以直接通过 Bean 对象调用对应接口定义的方法 ((InitializingBean) bean).afterPropertiesSet，这两种方法的实现效果是一样的。

除了在初始化过程中完成的操作，destroy-method 注解和 DisposableBean 接口的定义都会在 Bean 对象初始化阶段完成，将注册销毁方法的信息定义到 DefaultSingletonBeanRegistry 类的 disposableBeans 属性中，以便后续进行统一处理。

> 📝 **注意**：这里介绍一个适配器的使用，因为反射调用和接口直接调用是两种方式，所以需要使用适配器进行包装。下面的代码参考 DisposableBeanAdapter 的具体实现。关于销毁方法，需要在虚拟机执行关闭之前处理，这里需要用到一个注册钩子的功能，如 Runtime.getRuntime().addShutdownHook(new Thread(() -> System.out.println("close！")))，也可以使用 ApplicationContext.close 方法关闭容器。

7.3 初始化和销毁实现

1. 工程结构

```
spring-step-07
└── src
    ├── main
    │   └── java
    │       └── cn.bugstack.springframework
    │           └── beans
    │               ├── factory
    │               │   ├── config
    │               │   │   ├── AutowireCapableBeanFactory.java
    │               │   │   ├── BeanDefinition.java
    │               │   │   ├── BeanFactoryPostProcessor.java
    │               │   │   ├── BeanPostProcessor.java
    │               │   │   ├── BeanReference.java
    │               │   │   ├── ConfigurableBeanFactory.java
    │               │   │   └── SingletonBeanRegistry.java
    │               │   └── support
    │               │       ├── AbstractAutowireCapableBeanFactory.java
    │               │       ├── AbstractBeanDefinitionReader.java
    │               │       ├── AbstractBeanFactory.java
    │               │       ├── BeanDefinitionReader.java
    │               │       ├── BeanDefinitionRegistry.java
    │               │       ├── CglibSubclassingInstantiationStrategy.java
    │               │       ├── DefaultListableBeanFactory.java
    │               │       ├── DefaultSingletonBeanRegistry.java
    │               │       ├── DisposableBeanAdapter.java
    │               │       ├── InstantiationStrategy.java
    │               │       └── SimpleInstantiationStrategy.java
```

```
|   |   |   |   ├── xml
|   |   |   |   |   └── XmlBeanDefinitionReader.java
|   |   |   |   ├── BeanFactory.java
|   |   |   |   ├── ConfigurableListableBeanFactory.java
|   |   |   |   ├── DisposableBean.java
|   |   |   |   ├── HierarchicalBeanFactory.java
|   |   |   |   ├── InitializingBean.java
|   |   |   |   └── ListableBeanFactory.java
|   |   |   ├── BeansException.java
|   |   |   ├── PropertyValue.java
|   |   |   └── PropertyValues.java
|   |   ├── context
|   |   |   ├── support
|   |   |   |   ├── AbstractApplicationContext.java
|   |   |   |   ├── AbstractRefreshableApplicationContext.java
|   |   |   |   ├── AbstractXmlApplicationContext.java
|   |   |   |   └── ClassPathXmlApplicationContext.java
|   |   |   ├── ApplicationContext.java
|   |   |   └── ConfigurableApplicationContext.java
|   |   ├── core.io
|   |   |   ├── ClassPathResource.java
|   |   |   ├── DefaultResourceLoader.java
|   |   |   ├── FileSystemResource.java
|   |   |   ├── Resource.java
|   |   |   ├── ResourceLoader.java
|   |   |   └── UrlResource.java
|   |   └── utils
|   |       └── ClassUtils.java
└── test
    └── java
        └── cn.bugstack.springframework.test
            ├── bean
            |   ├── UserDao.java
            |   └── UserService.java
            └── ApiTest.java
```

Spring 应用上下文和对 Bean 对象扩展机制的类的关系如图 7-3 所示。

整个类图的结构是本次新增 Bean 实例化过程中的初始化方法和销毁方法。

因为一共实现了两种方式的初始化和销毁方法——XML 配置和定义接口——所以这里既包括 InitializingBean 接口、DisposableBean 接口，也需要使用 XmlBeanDefinitionReader 类将 spring.xml 配置文件的信息加载到 BeanDefinition 中。

第 7 章 Bean 对象的初始化和销毁

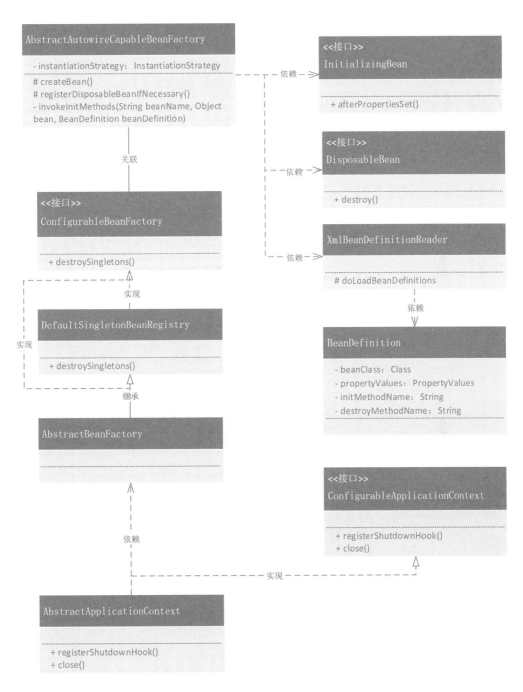

图 7-3

ConfigurableBeanFactory 接口定义了 destroySingletons 销毁方法，并由 AbstractBeanFactory 继承的父类 DefaultSingletonBeanRegistry 实现 ConfigurableBeanFactory 接口定义的 destroySingletons 方法。

> **注意**：虽然大多数程序员没有使用这种设计方式，一般都采用谁实现接口谁完成实现类，而不是把实现接口又交给继承的父类处理。但这种方式是一种不错的隔离服务分层的设计技巧，也可以在一些复制的业务场景中使用。

关于向虚拟机注册钩子，需要在关闭虚拟机之前销毁，即 Runtime.getRuntime().addShutdownHook(new Thread(() -> System.out.println("close！")))。

2. 定义初始化和销毁方法的接口

源码详见：cn.bugstack.springframework.beans.factory.InitializingBean。

```java
public interface InitializingBean {

    /**
     * 在 Bean 对象属性填充完成后调用
     *
     * @throws Exception
     */
    void afterPropertiesSet() throws Exception;

}
```

源码详见：cn.bugstack.springframework.beans.factory.DisposableBean。

```java
public interface DisposableBean {

    void destroy() throws Exception;

}
```

InitializingBean 和 DisposableBean 两个接口方法都是比较常用的。在一些需要结合 Spring 实现的组件中，我们经常使用这两个接口方法对参数进行初始化和销毁，如接口暴露、数据库数据读取、配置文件加载等。

3. Bean 属性定义新增初始化和销毁

源码详见：cn.bugstack.springframework.beans.factory.config.BeanDefinition。

```java
public class BeanDefinition {

    private Class beanClass;

    private PropertyValues propertyValues;

    private String initMethodName;

    private String destroyMethodName;

    // ...get/set
}
```

在 BeanDefinition 中新增了两个属性——initMethodName、destroyMethodName，目的是在 spring.xml 配置文件的 Bean 对象中可以配置 init-method="initDataMethod" destroy-method="destroyDataMethod"，最终实现接口的效果是一样的。只不过一种是直接调用接口方法，另一种是在配置文件中读取方法反射调用。

4．执行 Bean 对象的初始化方法

源码详见：cn.bugstack.springframework.beans.factory.support.AbstractAutowireCapableBeanFactory。

```java
public abstract class AbstractAutowireCapableBeanFactory extends AbstractBeanFactory
implements AutowireCapableBeanFactory {

    private InstantiationStrategy instantiationStrategy = new CglibSubclassingInstantiationStrategy();

    @Override
    protected Object createBean(String beanName, BeanDefinition beanDefinition, Object[] args) throws BeansException {
        Object bean = null;
        try {
            bean = createBeanInstance(beanDefinition, beanName, args);
            // 给 Bean 对象填充属性
            applyPropertyValues(beanName, bean, beanDefinition);
            // 执行 Bean 对象的初始化方法和 BeanPostProcessor 接口的前置和后置处理方法
            bean = initializeBean(beanName, bean, beanDefinition);
        } catch (Exception e) {
            throw new BeansException("Instantiation of bean failed", e);
        }

        // ...
```

```java
        registerSingleton(beanName, bean);
        return bean;
    }

    private Object initializeBean(String beanName, Object bean, BeanDefinition beanDefinition) {
        // 1. 执行 BeanPostProcessor Before 前置处理
        Object wrappedBean = applyBeanPostProcessorsBeforeInitialization(bean, beanName);

        // 执行 Bean 对象的初始化方法
        try {
            invokeInitMethods(beanName, wrappedBean, beanDefinition);
        } catch (Exception e) {
            throw new BeansException("Invocation of init method of bean[" + beanName + "] failed", e);
        }

        // 2. 执行 BeanPostProcessor After 后置处理
        wrappedBean = applyBeanPostProcessorsAfterInitialization(bean, beanName);
        return wrappedBean;
    }

    private void invokeInitMethods(String beanName, Object bean, BeanDefinition beanDefinition) throws Exception {
        // 1. 实现 InitializingBean 接口
        if (bean instanceof InitializingBean) {
            ((InitializingBean) bean).afterPropertiesSet();
        }

        // 2. 配置信息 init-method {判断是为了避免二次销毁}
        String initMethodName = beanDefinition.getInitMethodName();
        if (StrUtil.isNotEmpty(initMethodName)) {
            Method initMethod = beanDefinition.getBeanClass().getMethod(initMethodName);
            if (null == initMethod) {
                throw new BeansException("Could not find an init method named '" + initMethodName + "' on bean with name '" + beanName + "'");
            }
            initMethod.invoke(bean);
        }
    }

}
```

AbstractAutowireCapableBeanFactory 抽象类中的 createBean 是用来创建 Bean 对象的

方法。在这个方法中,我们已经扩展了 BeanFactoryPostProcessor、BeanPostProcessor,这里继续完善执行 Bean 对象的初始化方法。

在 invokeInitMethods 方法中,主要分为两部分实现 InitializingBean 接口,以及处理 afterPropertiesSet 方法。然后判断配置信息 init-method 是否存在,执行反射调用 initMethod.invoke(bean)。

在 Bean 对象初始化过程中,使用 createBean 和 invokeInitMethods 这两种方法都可以处理加载 Bean 对象中的初始化,使用户可以扩展额外的处理逻辑。

5. 定义销毁方法适配器(接口和配置)

源码详见:cn.bugstack.springframework.beans.factory.support.DisposableBeanAdapter。

```java
public class DisposableBeanAdapter implements DisposableBean {

    private final Object bean;
    private final String beanName;
    private String destroyMethodName;

    public DisposableBeanAdapter(Object bean, String beanName, BeanDefinition beanDefinition) {
        this.bean = bean;
        this.beanName = beanName;
        this.destroyMethodName = beanDefinition.getDestroyMethodName();
    }

    @Override
    public void destroy() throws Exception {
        // 1. 实现 DisposableBean 接口
        if (bean instanceof DisposableBean) {
            ((DisposableBean) bean).destroy();
        }

        // 2. 配置信息 destroy-method { 判断是为了避免二次销毁 }
        if (StrUtil.isNotEmpty(destroyMethodName) && !(bean instanceof DisposableBean && "destroy".equals(this.destroyMethodName))) {
            Method destroyMethod = bean.getClass().getMethod(destroyMethodName);
            if (null == destroyMethod) {
                throw new BeansException("Couldn't find a destroy method named '" + destroyMethodName + "' on bean with name '" + beanName + "'");
            }
            destroyMethod.invoke(bean);
        }
```

        }
    }

使用适配器的类的原因是，销毁方法有两种甚至多种方式，目前有实现接口 DisposableBean 和配置信息 destroy-method 两种方式，而这两种方式的销毁是由 Abstract ApplicationContext 向虚拟机注册钩子后、虚拟机关闭前执行的。

因为在销毁时，更希望有统一的接口进行销毁，所以这里新增了适配类，进行统一处理。

6. 创建 Bean 对象时注册销毁方法

源码详见：cn.bugstack.springframework.beans.factory.support.AbstractAutowireCapable BeanFactory。

```java
public abstract class AbstractAutowireCapableBeanFactory extends AbstractBeanFactory implements AutowireCapableBeanFactory {

    private InstantiationStrategy instantiationStrategy = new CglibSubclassingInstantiationStrategy();

    @Override
    protected Object createBean(String beanName, BeanDefinition beanDefinition, Object[] args) throws BeansException {
        Object bean = null;
        try {
            bean = createBeanInstance(beanDefinition, beanName, args);
            // 给 Bean 对象填充属性
            applyPropertyValues(beanName, bean, beanDefinition);
            // 执行 Bean 对象的初始化方法和 BeanPostProcessor 接口的前置和后置处理方法
            bean = initializeBean(beanName, bean, beanDefinition);
        } catch (Exception e) {
            throw new BeansException("Instantiation of bean failed", e);
        }

        // 注册实现了 DisposableBean 接口的 Bean 对象
        registerDisposableBeanIfNecessary(beanName, bean, beanDefinition);

        registerSingleton(beanName, bean);
        return bean;
    }
```

```java
protected void registerDisposableBeanIfNecessary(String beanName, Object bean, 
BeanDefinition beanDefinition) {
    if (bean instanceof DisposableBean || StrUtil.isNotEmpty(beanDefinition.
getDestroyMethodName())) {
        registerDisposableBean(beanName, new DisposableBeanAdapter(bean, beanName, 
beanDefinition));
    }
}
```

在创建 Bean 对象的实例时，需要将销毁方法保存起来，方便销毁时调用。

DisposableBean 销毁方法的具体信息会被注册到 DefaultSingletonBeanRegistry 中新增的 Map disposableBeans 属性中，因为 DisposableBean 接口的方法最终可以被 AbstractApplicationContext 类的 close 方法通过 getBeanFactory().destroySingletons 来调用。

源码详见：cn.bugstack.springframework.beans.factory.support.DisposableBeanAdapter。

```java
public class DisposableBeanAdapter implements DisposableBean {

    @Override
    public void destroy() throws Exception {
        // 1. 实现 DisposableBean 接口
        if (bean instanceof DisposableBean) {
            ((DisposableBean) bean).destroy();
        }

        // 2. 注解配置 destroy-method { 判断是为了避免二次销毁 }
        if (StrUtil.isNotEmpty(destroyMethodName) && !(bean instanceof DisposableBean 
&& "destroy".equals(this.destroyMethodName))) {
            Method destroyMethod = bean.getClass().getMethod(destroyMethodName);
            if (null == destroyMethod) {
                throw new BeansException("Couldn't find a destroy method named '" + 
destroyMethodName + "' on bean with name '" + beanName + "'");
            }
            destroyMethod.invoke(bean);
        }

    }

}
```

在使用 registerDisposableBeanIfNecessary 注册销毁方法时，会根据接口类型和配置类型统一交给 DisposableBeanAdapter 销毁适配器类进行统一处理。当通过 DisposableBeanAdapter#destroy 执行销毁方法时，会使用 Java 的关键字 instanceof 判断对象类型是否为 DisposableBean 并调用具体的销毁方法，以及使用反射调用处理 XML 配置的方法销毁。

7. 注册、关闭虚拟机钩子的方法

源码详见：cn.bugstack.springframework.context.ConfigurableApplicationContext。

```
public interface ConfigurableApplicationContext extends ApplicationContext {

    void refresh() throws BeansException;

    void registerShutdownHook();

    void close();

}
```

在 ConfigurableApplicationContext 接口中定义注册虚拟机钩子的 registerShutdownHook 方法和手动执行关闭的 close 方法。

源码详见：cn.bugstack.springframework.context.support.AbstractApplicationContext。

```
public abstract class AbstractApplicationContext extends DefaultResourceLoader
implements ConfigurableApplicationContext {

    // ...

    @Override
    public void registerShutdownHook() {
        Runtime.getRuntime().addShutdownHook(new Thread(this::close));
    }

    @Override
    public void close() {
        getBeanFactory().destroySingletons();
    }

}
```

这里的主要内容是如何实现注册和关闭虚拟机钩子的方法。用户可以尝试验证上文提到过的 Runtime.getRuntime().addShutdownHook 方法。在一些中间件和监控系统的设计中也可以使用这个方法，如监测服务器宕机、备机启动。

7.4 容器功能测试

1. 事先准备

源码详见：cn.bugstack.springframework.test.bean.UserDao。

```java
public class UserDao {

    private static Map<String, String> hashMap = new HashMap<>();

    public void initDataMethod(){
        System.out.println("执行: init-method");
        hashMap.put("10001", "小傅哥");
        hashMap.put("10002", "八杯水");
        hashMap.put("10003", "阿毛");
    }

    public void destroyDataMethod(){
        System.out.println("执行: destroy-method");
        hashMap.clear();
    }

    public String queryUserName(String uId) {
        return hashMap.get(uId);
    }

}
```

源码详见：cn.bugstack.springframework.test.bean.UserService。

```java
public class UserService implements InitializingBean, DisposableBean {

    private String uId;
    private String company;
    private String location;
    private UserDao userDao;
```

```java
    @Override
    public void destroy() throws Exception {
        System.out.println("执行: UserService.destroy");
    }

    @Override
    public void afterPropertiesSet() throws Exception {
        System.out.println("执行: UserService.afterPropertiesSet");
    }

    // ...get/set
}
```

UserDao 将之前使用 static 静态块初始化数据的方式，修改为使用 initDataMethod 和 destroyDataMethod 两个更简便的方式进行处理。

使用 UserService 来实现 InitializingBean 接口、DisposableBean 接口的 destroy 和 afterPropertiesSet 两个方法，并进行相应的初始化和销毁。

2. 配置文件

基础配置文件。无 BeanFactoryPostProcessor 和 BeanPostProcessor 实现类。

```xml
<?xml version="1.0" encoding="UTF-8"?>
<beans>

    <bean id="userDao" class="cn.bugstack.springframework.test.bean.UserDao" init-method="initDataMethod" destroy-method="destroyDataMethod"/>

    <bean id="userService" class="cn.bugstack.springframework.test.bean.UserService">
        <property name="uId" value="10001"/>
        <property name="company" value="腾讯"/>
        <property name="location" value="深圳"/>
        <property name="userDao" ref="userDao"/>
    </bean>

</beans>
```

配置文件中主要新增了 init-method="initDataMethod"、destroy-method="destroyDataMethod" 两个配置。从源码中可以知道，这两个配置是为了加入 BeanDefinition 定义类之后，写入 DefaultListableBeanFactory 类的 beanDefinitionMap 属性中。

3. 单元测试

```
@Test
public void test_xml() {
    // 1. 初始化 BeanFactory 接口
    ClassPathXmlApplicationContext applicationContext = new ClassPathXmlApplicationContext
("classpath:spring.xml");
    applicationContext.registerShutdownHook();

    // 2. 获取 Bean 对象的调用方法
    UserService userService = applicationContext.getBean("userService", UserService.class);
    String result = userService.queryUserInfo();
    System.out.println("测试结果: " + result);
}
```

在测试方法中增加了一个注册钩子 applicationContext.registerShutdownHook。

测试结果如下。

```
执行: init-method
执行: UserService.afterPropertiesSet
测试结果: 小傅哥，腾讯，深圳
执行: UserService.destroy
执行: destroy-method

Process finished with exit code 0
```

从测试结果中可以看到，使用新增的初始化方法和销毁方法已经可以正常输出结果了。

7.5 本章总结

本章主要实现通过 Spring 框架对外提供的 InitializingBean 接口和 DisposableBean 接口，以及通过 spring.xml 配置文件增设 init-method="initDataMethod" 配置和 destroy-method= "destroyDataMethod" 配置的方式，处理 Bean 对象在整个生命周期中的初始化和销毁。

通过学习本章的内容，读者可以看到目前 Spring 框架对 Bean 的功能越来越完善了，可扩展性也在不断增强。我们既可以在 Bean 注册完成实例化之前进行 BeanFactory

PostProcessor 操作，也可以在 Bean 实例化过程中执行前置和后置操作，还可以执行 Bean 对象的初始化方法和销毁方法。所以，一个简单的 Bean 对象已经被不断地赋予了各种扩展功能。

读者在学习和动手实践 Spring 框架的过程中，特别要注意对接口和抽象类的把握和使用，尤其当遇到类似 A 继承 B 实现 C 时，C 的接口方法由 A 继承的父类 B 实现。也可以将其应用到业务系统的开发中，处理一些复杂逻辑的功能分层，增加程序的可扩展、易维护等特性。

第 8 章
感知容器对象

当我们不断地深入学习 Spring 框架后,可以感受到它带来的强大的扩展功能。例如,可以在对象加载完成未初始化时修改对象信息,也可以在对象初始化前后修改属性,还可以获取整个 Bean 对象生命周期中的名称、容器和上下文等信息。

这些强大的扩展功能都来自工程架构和设计原则的运用、解耦系统、降低耦合。本章仍会通过依赖倒置实现开闭原则的方式,提供感知容器变化的功能,允许用户扩展满足自己的功能需求。

- 本章难度:★★★☆☆
- 本章重点:通过依赖倒置定义感知容器的标记接口,使 Bean 对象生命周期中的节点作为接口实现,并在对象实例化后的初始化阶段进行调用,将信息传递给用户。

8.1 Spring Bean 容器的功能

目前已实现的 Spring 框架,在 Bean 对象方面能提供的功能包括:Bean 对象的定义和注册,在 Bean 对象过程中执行 BeanFactoryPostProcessor、BeanPostProcessor、InitializingBean、DisposableBean 操作,以及在 XML 中新增初始化和销毁的配置处理,使得 Bean 对象具有更强的操作性。

如果想要对 Spring 框架提供的 BeanFactory、ApplicationContext、BeanClassLoader 等

功能进行功能扩展，就可以在 Spring 框架中提供一种能感知容器的接口。如果实现了这个接口，就可以获取接口入参信息中各类对象提供的功能，并进行一些额外逻辑的处理。

8.2 感知容器设计

如果想要获取 Spring 框架提供的资源，则要考虑以什么方式获取。对于定义好的获取方式，在 Spring 框架中应该怎样承接。一旦实现这两项内容，就可以扩展出一些属于 Spring 框架本身的功能。

在 Bean 对象的实例化阶段，我们进行了额外定义、属性、初始化和销毁等，如果想获取 Spring 框架中的 BeanFactory 接口、ApplicationContext 接口，也可以通过此种设计方法获取。此时需要定义一个标记性的接口，这个接口不需要使用方法，只起到标记作用。该接口的具体功能由继承此接口的其他功能性接口定义，此时可以通过 Java 关键字 instanceof 判断和调用了。整体设计结构如图 8-1 所示。

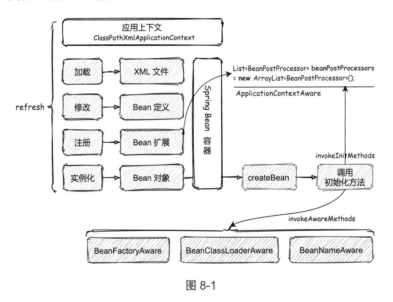

图 8-1

定义 Aware 接口。在 Spring 框架中，Aware 是一种感知标记性接口，子类的定义和实现能感知容器中的相关对象。即通过 Aware 接口，可以向具体的实现类中提供容器服务。

继承 Aware 的接口包括 BeanFactoryAware、BeanClassLoaderAware、BeanNameAware

和 ApplicationContextAware。在 Spring 源码中，还有一些其他关于注解的接口，不过目前还使用不到。

在接口具体的实现过程中可以看到，一部分接口（BeanFactoryAware、BeanClassLoaderAware、BeanNameAware）在 factory 的 support 文件夹中，其余的接口（ApplicationContextAware）在 context 的 support 文件夹中，因为获取不同的内容需要在不同的包下提供，所以，在 AbstractApplicationContext 的具体实现中会使用向 beanFactory 中添加 BeanPostProcessor 内容的 ApplicationContextAwareProcessor 操作，最后由 AbstractAutowireCapableBeanFactory 创建 createBean 时进行相应的调用。

8.3 感知容器实现

1. 工程结构

```
spring-step-08
└── src
    ├── main
    │   └── java
    │       └── cn.bugstack.springframework
    │           ├── beans
    │           │   ├── factory
    │           │   │   ├── config
    │           │   │   │   ├── AutowireCapableBeanFactory.java
    │           │   │   │   ├── BeanDefinition.java
    │           │   │   │   ├── BeanFactoryPostProcessor.java
    │           │   │   │   ├── BeanPostProcessor.java
    │           │   │   │   ├── BeanReference.java
    │           │   │   │   ├── ConfigurableBeanFactory.java
    │           │   │   │   └── SingletonBeanRegistry.java
    │           │   │   ├── support
    │           │   │   │   ├── AbstractAutowireCapableBeanFactory.java
    │           │   │   │   ├── AbstractBeanDefinitionReader.java
    │           │   │   │   ├── AbstractBeanFactory.java
    │           │   │   │   ├── BeanDefinitionReader.java
    │           │   │   │   ├── BeanDefinitionRegistry.java
    │           │   │   │   ├── CglibSubclassingInstantiationStrategy.java
    │           │   │   │   ├── DefaultListableBeanFactory.java
    │           │   │   │   ├── DefaultSingletonBeanRegistry.java
    │           │   │   │   ├── DisposableBeanAdapter.java
```

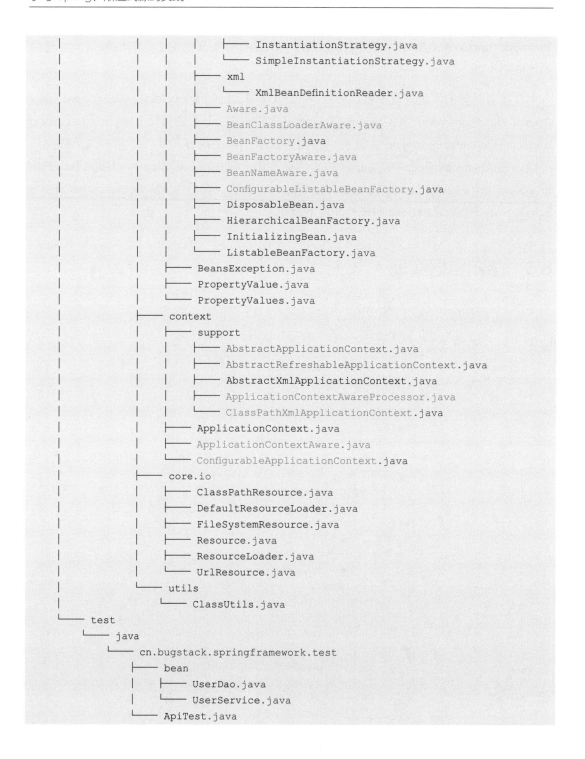

Spring 感知接口的设计和实现类的关系如图 8-2 所示。

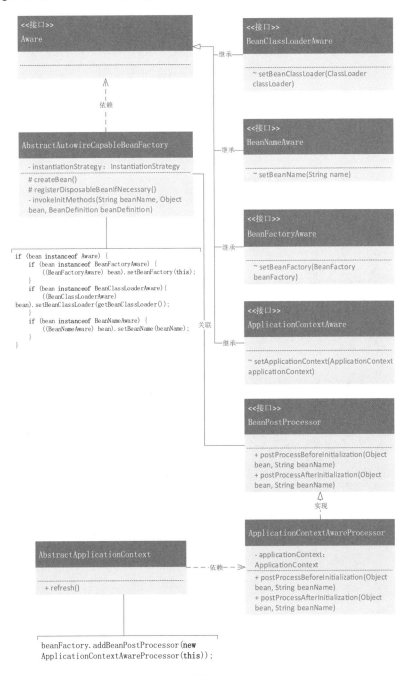

图 8-2

以上的整个类关系图就是关于感知的定义和对容器感知的实现。

Aware 有 4 个用于继承的接口，其他继承接口的目的是继承一个标记，有了标记，就可以更加方便地判断类的实现。

由于 ApplicationContext 接口并不是在 AbstractAutowireCapableBeanFactory 中 createBean 方法下的内容，所以需要向容器中注册 addBeanPostProcessor，当由 createBean 统一调用 applyBeanPostProcessorsBeforeInitialization 时执行。

2. 定义标记接口

源码详见：cn.bugstack.springframework.beans.factory.Aware。

```
/**
 * Marker superinterface indicating that a bean is eligible to be
 * notified by the Spring container of a particular framework object
 * through a callback-style method. Actual method signature is
 * determined by individual subinterfaces, but should typically
 * consist of just one void-returning method that accepts a single
 * argument
 *
 * 标记类接口，实现该接口后可以被 Spring Bean 容器感知
 *
 */
public interface Aware {
}
```

在 Spring 中，有很多类似这种标记接口的设计方式，它们的存在就像是一种标签，便于统一获取出属于此类接口的实现类。

3. 容器感知类

（1）BeanFactoryAware。

源码详见：cn.bugstack.springframework.beans.factory.BeanFactoryAware。

```
public interface BeanFactoryAware extends Aware {

    void setBeanFactory(BeanFactory beanFactory) throws BeansException;

}
```

源码注释 Interface to be implemented by beans that wish to be aware of their owning {@link BeanFactory}，意为：实现此接口，可以感知到所属的 BeanFactory。

（2）BeanClassLoaderAware。

源码详见：cn.bugstack.springframework.beans.factory.BeanClassLoaderAware。

```
public interface BeanClassLoaderAware extends Aware{

    void setBeanClassLoader(ClassLoader classLoader);

}
```

源码注释 Callback that allows a bean to be aware of the bean{@link ClassLoader class loader}; that is, the class loader used by the present bean factory to load bean classes，意为：实现此接口，可以感知到所属的 ClassLoader 类加载器。

（3）BeanNameAware。

源码详见：cn.bugstack.springframework.beans.factory.BeanNameAware。

```
public interface BeanNameAware extends Aware {

    void setBeanName(String name);

}
```

源码注释 Interface to be implemented by beans that want to be aware of their bean name in a bean factory，意为：实现此接口，可以感知到所属的 Bean 对象的名称。

（4）ApplicationContextAware。

源码详见：cn.bugstack.springframework.context.ApplicationContextAware。

```
public interface ApplicationContextAware extends Aware {

    void setApplicationContext(ApplicationContext applicationContext) throws BeansException;

}
```

源码注释 Interface to be implemented by any object that wishes to be notifiedof the {@link ApplicationContext} that it runs in，意为：实现此接口，可以感知到所属的 ApplicationContext 应用上下文信息。

4. 包装处理器（ApplicationContextAwareProcessor）

源码详见：cn.bugstack.springframework.context.support.ApplicationContextAwareProcessor。

```java
public class ApplicationContextAwareProcessor implements BeanPostProcessor {

    private final ApplicationContext applicationContext;

    public ApplicationContextAwareProcessor(ApplicationContext applicationContext) {
        this.applicationContext = applicationContext;
    }

    @Override
    public Object postProcessBeforeInitialization(Object bean, String beanName) throws BeansException {
        if (bean instanceof ApplicationContextAware){
            ((ApplicationContextAware) bean).setApplicationContext(applicationContext);
        }
        return bean;
    }

    @Override
    public Object postProcessAfterInitialization(Object bean, String beanName) throws BeansException {
        return bean;
    }

}
```

由于并不能直接在创建 Bean 时获取 ApplicationContext 属性，所以需要在执行 refresh 时，将 ApplicationContext 写入一个包装的 BeanPostProcessor 类中，再使用 AbstractAutowireCapableBeanFactory.applyBeanPostProcessorsBeforeInitialization 方法调用时获取 ApplicationContext 属性。

5. 注册 BeanPostProcessor

源码详见：cn.bugstack.springframework.context.support.AbstractApplicationContext。

```java
public abstract class AbstractApplicationContext extends DefaultResourceLoader implements ConfigurableApplicationContext {

    @Override
    public void refresh() throws BeansException {
        // 1. 创建 BeanFactory，并加载 BeanDefinition
        refreshBeanFactory();

        // 2. 获取 BeanFactory
        ConfigurableListableBeanFactory beanFactory = getBeanFactory();
```

```
    // 3. 添加 ApplicationContextAwareProcessor 类，让继承自 ApplicationContextAware
    // 接口的 Bean 对象都能感知所属的 ApplicationContext
    beanFactory.addBeanPostProcessor(new ApplicationContextAwareProcessor(this));

    // 4. 在将 Bean 对象实例化之前，执行 BeanFactoryPostProcessor 操作
    invokeBeanFactoryPostProcessors(beanFactory);

    // 5. BeanPostProcessor 需要在其他 Bean 对象实例化之前注册
    registerBeanPostProcessors(beanFactory);

    // 6. 提前实例化单例 Bean 对象
    beanFactory.preInstantiateSingletons();
}

// ...
}
```

- refresh 方法就是整个 Spring Bean 容器的操作过程，与前面章节对比，本次增加了关于 addBeanPostProcessor 的操作。

- 添加 ApplicationContextAwareProcessor 类，让继承自 ApplicationContextAware 接口的 Bean 对象都能感知所属的 ApplicationContext。

6. 感知调用

源码详见：cn.bugstack.springframework.beans.factory.support.AbstractAutowireCapableBeanFactory。

```
public abstract class AbstractAutowireCapableBeanFactory extends AbstractBeanFactory
implements AutowireCapableBeanFactory {

    private InstantiationStrategy instantiationStrategy = new CglibSubclassingInstantiationStrategy();

    @Override
    protected Object createBean(String beanName, BeanDefinition beanDefinition,
Object[] args) throws BeansException {
        Object bean = null;
        try {
            bean = createBeanInstance(beanDefinition, beanName, args);
            // 给 Bean 对象填充属性
            applyPropertyValues(beanName, bean, beanDefinition);
            // 执行 Bean 对象的初始化方法和 BeanPostProcessor 接口的前置处理方法与后置处理方法
            bean = initializeBean(beanName, bean, beanDefinition);
```

```java
        } catch (Exception e) {
            throw new BeansException("Instantiation of bean failed", e);
        }

        // 注册实现了 DisposableBean 接口的 Bean 对象
        registerDisposableBeanIfNecessary(beanName, bean, beanDefinition);

        registerSingleton(beanName, bean);
        return bean;
    }

    private Object initializeBean(String beanName, Object bean, BeanDefinition beanDefinition) {

        // invokeAwareMethods
        if (bean instanceof Aware) {
            if (bean instanceof BeanFactoryAware) {
                ((BeanFactoryAware) bean).setBeanFactory(this);
            }
            if (bean instanceof BeanClassLoaderAware){
                ((BeanClassLoaderAware) bean).setBeanClassLoader(getBeanClassLoader());
            }
            if (bean instanceof BeanNameAware) {
                ((BeanNameAware) bean).setBeanName(beanName);
            }
        }

        // 执行 BeanPostProcessor Before 前置处理
        Object wrappedBean = applyBeanPostProcessorsBeforeInitialization(bean, beanName);

        // 执行 Bean 对象的初始化方法
        try {
            invokeInitMethods(beanName, wrappedBean, beanDefinition);
        } catch (Exception e) {
            throw new BeansException("Invocation of init method of bean[" + beanName + "] failed", e);
        }

        // 执行 BeanPostProcessor After 后置处理
        wrappedBean = applyBeanPostProcessorsAfterInitialization(bean, beanName);
        return wrappedBean;
    }

    @Override
```

```java
    public Object applyBeanPostProcessorsBeforeInitialization(Object existingBean, 
String beanName) throws BeansException {
        Object result = existingBean;
        for (BeanPostProcessor processor : getBeanPostProcessors()) {
            Object current = processor.postProcessBeforeInitialization(result, beanName);
            if (null == current) return result;
            result = current;
        }
        return result;
    }

    @Override
    public Object applyBeanPostProcessorsAfterInitialization(Object existingBean, 
String beanName) throws BeansException {
        Object result = existingBean;
        for (BeanPostProcessor processor : getBeanPostProcessors()) {
            Object current = processor.postProcessAfterInitialization(result, beanName);
            if (null == current) return result;
            result = current;
        }
        return result;
    }

}
```

> 注意：这里省略了一些类的内容，只保留本次 Aware 接口。

在 initializeBean 接口中，通过判断条件 bean instanceof Aware 调用了 3 个接口方法——BeanFactoryAware.setBeanFactory(this)、BeanClassLoaderAware.setBeanClassLoader(getBeanClassLoader())、BeanNameAware.setBeanName(beanName)，这样就能通知已经实现了 initializeBean 接口的类。

另外，还向 BeanPostProcessor 类中添加了 ApplicationContextAwareProcessor 方法，此时这个方法也会被调用到具体的类实现，得到一个 ApplicationContext 属性。

8.4 Aware 接口的功能测试

1. 事先准备

源码详见：cn.bugstack.springframework.test.bean.UserDao。

```java
public class UserDao {

    private static Map<String, String> hashMap = new HashMap<>();

    public void initDataMethod(){
        System.out.println("执行: init-method");
        hashMap.put("10001", "小傅哥");
        hashMap.put("10002", "八杯水");
        hashMap.put("10003", "阿毛");
    }

    public void destroyDataMethod(){
        System.out.println("执行: destroy-method");
        hashMap.clear();
    }

    public String queryUserName(String uId) {
        return hashMap.get(uId);
    }

}
```

源码详见：cn.bugstack.springframework.test.bean.UserService。

```java
public class UserService implements BeanNameAware, BeanClassLoaderAware, ApplicationContextAware, BeanFactoryAware {

    private ApplicationContext applicationContext;
    private BeanFactory beanFactory;

    private String uId;
    private String company;
    private String location;
    private UserDao userDao;

    @Override
    public void setBeanFactory(BeanFactory beanFactory) throws BeansException {
        this.beanFactory = beanFactory;
    }

    @Override
    public void setApplicationContext(ApplicationContext applicationContext) throws BeansException {
        this.applicationContext = applicationContext;
    }
```

```
    @Override
    public void setBeanName(String name) {
        System.out.println("Bean Name is: " + name);
    }

    @Override
    public void setBeanClassLoader(ClassLoader classLoader) {
        System.out.println("ClassLoader: " + classLoader);
    }

    // ...get/set
}
```

此次 UserDao 的功能并没有改变，还是提供了关于初始化的方法，并在 spring.xml 配置文件中提供 init-method、destroy-method 配置信息。

UserService 增加了 BeanNameAware、BeanClassLoaderAware、ApplicationContextAware、BeanFactoryAware 共 4 个感知的实现类，并在类中实现相应的接口方法。

2. 配置文件

基础配置文件。无 BeanFactoryPostProcessor、BeanPostProcessor 等实现类。

```
<?xml version="1.0" encoding="UTF-8"?>
<beans>

    <bean id="userDao" class="cn.bugstack.springframework.test.bean.UserDao" init-method="initDataMethod" destroy-method="destroyDataMethod"/>

    <bean id="userService" class="cn.bugstack.springframework.test.bean.UserService">
        <property name="uId" value="10001"/>
        <property name="company" value="腾讯"/>
        <property name="location" value="深圳"/>
        <property name="userDao" ref="userDao"/>
    </bean>

</beans>
```

本节并没有额外新增配置信息，与前面章节的内容相同。

3. 单元测试

```
@Test
public void test_xml() {
```

```
    // 1. 初始化 BeanFactory 接口
    ClassPathXmlApplicationContext applicationContext = new ClassPathXmlApplicationContext
("classpath:spring.xml");
    applicationContext.registerShutdownHook();

    // 2. 获取 Bean 对象的调用方法
    UserService userService = applicationContext.getBean("userService", UserService.class);
    String result = userService.queryUserInfo();
    System.out.println("测试结果: " + result);
    System.out.println("ApplicationContextAware: "+userService.getApplicationContext());
    System.out.println("BeanFactoryAware: "+userService.getBeanFactory());
}
```

在测试方法中主要添加了关于新增 Aware 实现的调用，其他不需要实现的调用也输出了相应的日志信息，可以在测试结果中看到。

测试结果如下。

```
执行: init-method
ClassLoader: sun.misc.Launcher$AppClassLoader@14dad5dc
Bean Name is: userService
测试结果: 小傅哥, 腾讯, 深圳
ApplicationContextAware: cn.bugstack.springframework.context.support.
ClassPathXmlApplicationContext@5ba23b66
BeanFactoryAware: cn.bugstack.springframework.beans.factory.support.
DefaultListableBeanFactory@2ff4f00f
执行: destroy-method

Process finished with exit code 0
```

从测试结果中可以看到，本节新增的感知接口对应的具体实现（BeanNameAware、BeanClassLoaderAware、ApplicationContextAware、BeanFactoryAware）已经可以正常输出结果了。

8.5 本章总结

关于 Spring 框架的实现，某些功能已经趋于完整，尤其是 Bean 对象的生命周期已经有了很多的体现，如图 8-3 所示。

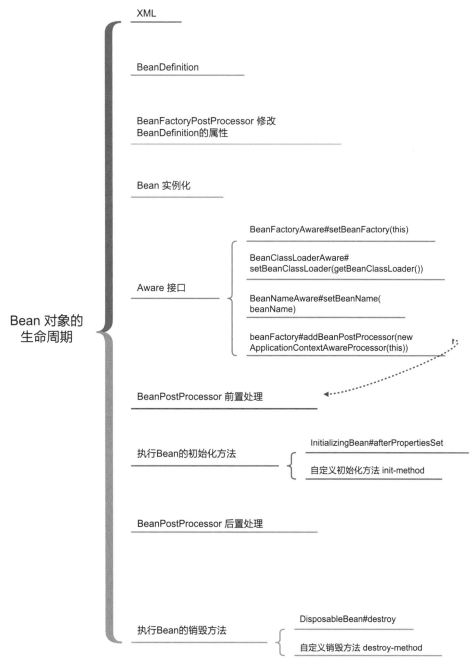

图 8-3

本章主要讲解了 Aware 接口的 4 个继承接口 BeanNameAware、BeanClassLoaderAware、ApplicationContextAware、BeanFactoryAware 的实现，以及扩展了 Spring 的功能。当开发 Spring 中间件时，会经常用到这些类。

每一节内容的实现都是在以设计模式为核心的结构上补充各项模块的功能，只编写代码并不会有太多收获。读者一定要理解为什么这样设计，这样设计的好处是什么，为什么有那么多接口和抽象类的应用，这些才是学习 Spring 框架的核心。

第 9 章
对象作用域和 FactoryBean

Spring 框架中的 BeanFactory 和 FactoryBean 有什么区别？这可能是大家在日常工作或者学习时常遇到的问题。

在实现 Spring 框架的功能时，BeanFactory 是 IOC 容器最基本的接口形式。从名字上可以知道，BeanFactory 是 Bean 的工厂，而 FactoryBean 是一个工厂对象。第三方对象生产者提供 FactoryBean 的实现类，包装出一个复杂的对象并忽略一些处理细节。本章主要介绍如何实现 FactoryBean 并演示最基本的代理反射思想的使用方法，这也是 ORM 的雏形。

- 本章难度：★★★☆☆
- 本章重点：提供 FactoryBean 接口的定义，使用户可以扩展创建复杂的代理 Bean 对象，将 Spring 与其他框架建立容器对象管理。

9.1 Bean 对象的来源和模式

在集合 Spring 框架下使用的 MyBatis 框架中，Bean 对象的核心作用是使用户无须实现 Dao 接口类，就可以通过 XML 或者注解配置的方式完成对数据库执行 CRUD 操作。那么在实现 ORM 框架的过程中，如何将一个数据库操作的 Bean 对象交给 Spring 管理呢？

在使用 MyBatis 框架时，通常不会手动创建任何操作数据库的 Bean 对象，仅有一个接口定义，而这个接口定义可以被注入其他需要使用 Dao 的属性中。这个过程需要解决的问题是，如何将复杂且以代理方式动态变化的对象注册到 Spring Bean 容器中。

9.2 FactoryBean 和对象模式设计

提供一个能让用户定义的复杂的 Bean 对象，其意义非常大，这样 Spring 的生态种子孵化箱就诞生了，任何框架都可以在此标准上接入相应的服务。

整个 Spring 框架在开发过程中就已经提供了各项扩展功能的接口，只需在合适的位置提供一个处理接口调用和相应的功能逻辑。本节的目标是实现对外提供一个可以从 FactoryBean 的 getObject 方法中二次获取对象的功能，使所有实现此接口的对象类都可以扩充自己的对象功能。MyBatis 就实现了一个 MapperFactoryBean 类，在 getObject 方法中提供了 SqlSession 接口并执行 CRUD 操作。整体设计结构如图 9-1 所示。

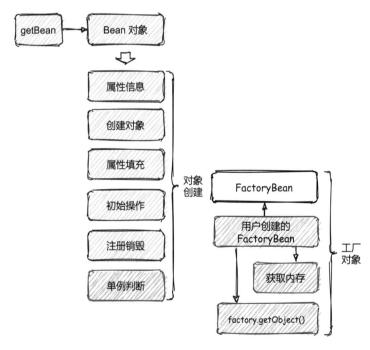

图 9-1

整个实现过程包括两部分，一部分是实现单例或原型对象，另一部分是实现通过

FactoryBean 获取具体调用对象的 getObject 操作。

SCOPE_SINGLETON 和 SCOPE_PROTOTYPE 对象的创建与获取方式的主要区别在于，在创建完 AbstractAutowireCapableBeanFactory#createBean 对象后是否将其存储到内存中，如果没有存储到内存中，则每次获取对象时都会重新创建对象。

通过 createBean 执行完对象创建、属性填充、依赖加载、前置后置处理、初始化等操作后，就要开始判断整个对象是否为一个 FactoryBean。如果是，则继续执行获取 FactoryBean 具体对象中的 getObject 操作。在获取对象的整个过程中会新增一个单例类型的判断语句 factory.isSingleton，用于决定是否使用内存来存储对象信息。

9.3　FactoryBean 和对象模式实现

1. 工程结构

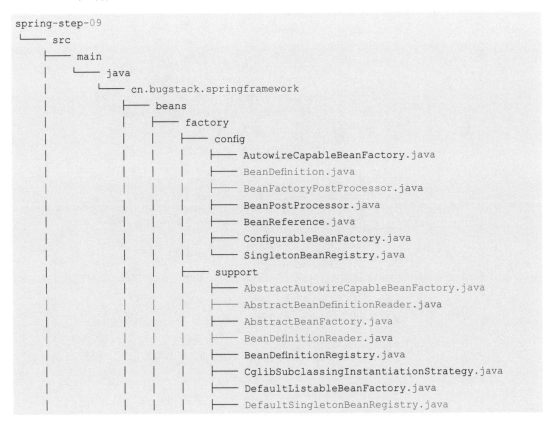

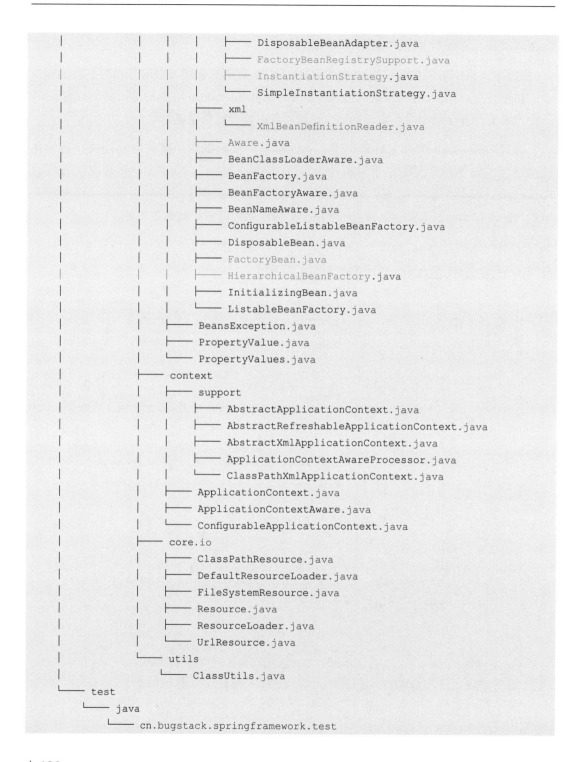

```
├── bean
│   ├── UserDao.java
│   │   └── UserService.java
└── ApiTest.java
```

Spring 单例、原型模式及 FactoryBean 功能的类关系如图 9-2 所示。

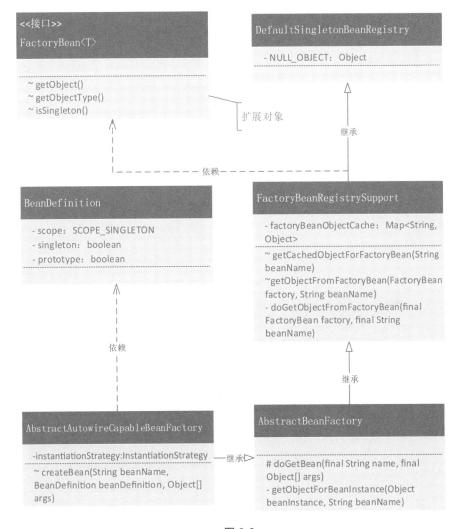

图 9-2

整个实现过程并不复杂，只是在现有的 AbstractAutowireCapableBeanFactory 类与继承的抽象类 AbstractBeanFactory 中进行扩展。

不过这次在 AbstractBeanFactory 继承的 DefaultSingletonBeanRegistry 类的中间加了一个 FactoryBeanRegistrySupport 类。它主要为 FactoryBean 注册提供了支撑。

2．Bean 对象的作用范围和 XML 解析

源码详见：cn.bugstack.springframework.beans.factory.config.BeanDefinition。

```java
public class BeanDefinition {

    String SCOPE_SINGLETON = ConfigurableBeanFactory.SCOPE_SINGLETON;

    String SCOPE_PROTOTYPE = ConfigurableBeanFactory.SCOPE_PROTOTYPE;

    private Class beanClass;

    private PropertyValues propertyValues;

    private String initMethodName;

    private String destroyMethodName;

    private String scope = SCOPE_SINGLETON;

    private boolean singleton = true;

    private boolean prototype = false;

    // …get/set
}
```

singleton、prototype 是在 BeanDefinition 类中新增加的两个属性，用于将从 spring.xml 配置文件中解析到的 Bean 对象的作用范围填充到属性中。

源码详见：cn.bugstack.springframework.beans.factory.xml.XmlBeanDefinitionReader。

```java
public class XmlBeanDefinitionReader extends AbstractBeanDefinitionReader {

    protected void doLoadBeanDefinitions(InputStream inputStream) throws ClassNotFoundException {

        for (int i = 0; i < childNodes.getLength(); i++) {
            // 判断元素
            if (!(childNodes.item(i) instanceof Element)) continue;
            // 判断对象
            if (!"bean".equals(childNodes.item(i).getNodeName())) continue;
```

```java
// 解析标签
Element bean = (Element) childNodes.item(i);
String id = bean.getAttribute("id");
String name = bean.getAttribute("name");
String className = bean.getAttribute("class");
String initMethod = bean.getAttribute("init-method");
String destroyMethodName = bean.getAttribute("destroy-method");
String beanScope = bean.getAttribute("scope");

// 获取 Class，方便获取类中的名称
Class<?> clazz = Class.forName(className);
// 判断优先级 id > name
String beanName = StrUtil.isNotEmpty(id) ? id : name;
if (StrUtil.isEmpty(beanName)) {
    beanName = StrUtil.lowerFirst(clazz.getSimpleName());
}

// 定义Bean对象
BeanDefinition beanDefinition = new BeanDefinition(clazz);
beanDefinition.setInitMethodName(initMethod);
beanDefinition.setDestroyMethodName(destroyMethodName);

if (StrUtil.isNotEmpty(beanScope)) {
    beanDefinition.setScope(beanScope);
}

// ...

// 注册 BeanDefinition
getRegistry().registerBeanDefinition(beanName, beanDefinition);
    }
}
}
```

在解析 XML 处理类 XmlBeanDefinitionReader 的过程中，新增了 Bean 对象配置中对 scope 的解析，并将这个属性填充到 Bean 的定义 beanDefinition.setScope(beanScope) 中。

3. 在创建和修改对象时，判断单例模式和原型模式

源码详见：cn.bugstack.springframework.beans.factory.support.AbstractAutowireCapableBeanFactory。

```java
public abstract class AbstractAutowireCapableBeanFactory extends AbstractBeanFactory
implements AutowireCapableBeanFactory {

    private InstantiationStrategy instantiationStrategy = new CglibSubclassingInstantiationStrategy();

    @Override
    protected Object createBean(String beanName, BeanDefinition beanDefinition,
Object[] args) throws BeansException {
        Object bean = null;
        try {
            bean = createBeanInstance(beanDefinition, beanName, args);
            // 给 Bean 对象填充属性
            applyPropertyValues(beanName, bean, beanDefinition);
            // 执行 Bean 对象的初始化方法和 BeanPostProcessor 接口的前置处理方法和后置处理方法
            bean = initializeBean(beanName, bean, beanDefinition);
        } catch (Exception e) {
            throw new BeansException("Instantiation of bean failed", e);
        }

        // 注册实现了 DisposableBean 接口的 Bean 对象
        registerDisposableBeanIfNecessary(beanName, bean, beanDefinition);

        // 判断 SCOPE_SINGLETON、SCOPE_PROTOTYPE
        if (beanDefinition.isSingleton()) {
            registerSingleton(beanName, bean);
        }
        return bean;
    }

    protected void registerDisposableBeanIfNecessary(String beanName, Object bean,
BeanDefinition beanDefinition) {
        // 非 Singleton 类型的 Bean 对象不必执行销毁方法
        if (!beanDefinition.isSingleton()) return;

        if (bean instanceof DisposableBean || StrUtil.isNotEmpty(beanDefinition.
getDestroyMethodName())) {
            registerDisposableBean(beanName, new DisposableBeanAdapter(bean, beanName,
beanDefinition));
        }
    }

    // 其他功能
}
```

单例模式（Singleton）和原型模式的区别为是否将 Bean 对象存储到内存中。如果是原型模式，就不会将 Bean 对象存储到内存中，每次获取都需要重新创建对象。非单例模式的 Bean 不需要执行销毁方法。所以这里的代码有两处修改，一处是在 createBean 中判断是否添加 registerSingleton(beanName, bean)，另一处是在 registerDisposableBeanIfNecessary 销毁注册中执行判断语句 if (!beanDefinition.isSingleton()) return。

4. 定义 FactoryBean 接口

源码详见：cn.bugstack.springframework.beans.factory.FactoryBean。

```java
public interface FactoryBean<T> {

    T getObject() throws Exception;

    Class<?> getObjectType();

    boolean isSingleton();

}
```

定义 FactoryBean 接口需要提供 3 个方法——获取对象、对象类型及是否为单例对象。如果是单例对象，将 Bean 对象存储到内存中。

5. 实现一个 FactoryBean 的注册服务

源码详见：cn.bugstack.springframework.beans.factory.support.FactoryBeanRegistrySupport。

```java
public abstract class FactoryBeanRegistrySupport extends DefaultSingletonBeanRegistry {

    /**
     * Cache of singleton objects created by FactoryBeans: FactoryBean name --> object
     */
    private final Map<String, Object> factoryBeanObjectCache = new ConcurrentHashMap<String, Object>();

    protected Object getCachedObjectForFactoryBean(String beanName) {
        Object object = this.factoryBeanObjectCache.get(beanName);
        return (object != NULL_OBJECT ? object : null);
    }

    protected Object getObjectFromFactoryBean(FactoryBean factory, String beanName) {
        if (factory.isSingleton()) {
            Object object = this.factoryBeanObjectCache.get(beanName);
```

```
            if (object == null) {
                object = doGetObjectFromFactoryBean(factory, beanName);
                this.factoryBeanObjectCache.put(beanName, (object != null ? object : NULL_OBJECT));
            }
            return (object != NULL_OBJECT ? object : null);
        } else {
            return doGetObjectFromFactoryBean(factory, beanName);
        }
    }

    private Object doGetObjectFromFactoryBean(final FactoryBean factory, final String beanName){
        try {
            return factory.getObject();
        } catch (Exception e) {
            throw new BeansException("FactoryBean threw exception on object[" + beanName + "] creation", e);
        }
    }
}
```

FactoryBeanRegistrySupport 类主要处理关于 FactoryBean 类对象的注册操作，之所以放到一个单独的类中，是为了不同领域模块下的类只负责各自需要完成的功能，避免因扩展导致类膨胀而难以维护。

同样，这里也定义了缓存操作 factoryBeanObjectCache，用于存储单例类型的对象，避免重复创建该对象。在日常使用中，也需要创建单例对象。

doGetObjectFromFactoryBean 是获取 FactoryBean#getObject 的方法，因为既要处理缓存又要获取对象，所以额外提供了 doGetObjectFromFactoryBean 方法进行逻辑包装，这部分的操作方式和日常的业务逻辑开发非常相似。如果无法从 Redis 中获取数据，则从其他数据库中获取数据并将其写入 Redis。

6. 扩展 AbstractBeanFactory 并创建对象逻辑

源码详见：cn.bugstack.springframework.beans.factory.support.AbstractBeanFactory。

```
public abstract class AbstractBeanFactory extends FactoryBeanRegistrySupport implements ConfigurableBeanFactory {
```

```java
protected <T> T doGetBean(final String name, final Object[] args) {
    Object sharedInstance = getSingleton(name);
    if (sharedInstance != null) {
        // 如果是 FactoryBean 类，则需要调用 FactoryBean#getObject
        return (T) getObjectForBeanInstance(sharedInstance, name);
    }

    BeanDefinition beanDefinition = getBeanDefinition(name);
    Object bean = createBean(name, beanDefinition, args);
    return (T) getObjectForBeanInstance(bean, name);
}

private Object getObjectForBeanInstance(Object beanInstance, String beanName) {
    if (!(beanInstance instanceof FactoryBean)) {
        return beanInstance;
    }

    Object object = getCachedObjectForFactoryBean(beanName);

    if (object == null) {
        FactoryBean<?> factoryBean = (FactoryBean<?>) beanInstance;
        object = getObjectFromFactoryBean(factoryBean, beanName);
    }

    return object;
}

// ...
}
```

这里将 AbstractBeanFactory 之前继承的 DefaultSingletonBeanRegistry 修改为 FactoryBeanRegistrySupport 为了扩展出创建 FactoryBean 的功能，需要在一个链条服务上截出一段来处理额外的服务，再将链条连接上。

这里新增的功能主要是在 doGetBean 方法中实现的，通过调用 (T) getObjectForBeanInstance(sharedInstance, name) 方法来获取 FactoryBean。

在 getObjectForBeanInstance 方法中执行具体的 instanceof 判断，并从 FactoryBean 的缓存中获取对象。如果 FactoryBean 缓存中不存在对象，则调用 FactoryBeanRegistrySupport#getObjectFromFactoryBean，并执行具体的操作。

9.4 代理 Bean 和对象模式测试

1. 事先准备

源码详见：cn.bugstack.springframework.test.bean.IUserDao。

```
public interface IUserDao {

    String queryUserName(String uId);

}
```

这里删除了 UserDao，定义了一个 IUserDao 接口，这样做是为了通过 FactoryBean 执行自定义对象的代理操作。

源码详见：cn.bugstack.springframework.test.bean.UserService。

```
public class UserService {

    private String uId;
    private String company;
    private String location;
    private IUserDao userDao;

    public String queryUserInfo() {
        return userDao.queryUserName(uId) + "," + company + "," + location;
    }

    // …get/set
}
```

在 UserService 中将原有的 UserDao 属性修改为 IUserDao，后面会给这个属性注册代理对象。

2. 定义 FactoryBean 接口

源码详见：cn.bugstack.springframework.test.bean.ProxyBeanFactory。

```
public class ProxyBeanFactory implements FactoryBean<IUserDao> {

    @Override
```

```java
    public IUserDao getObject() throws Exception {
        InvocationHandler handler = (proxy, method, args) -> {

            Map<String, String> hashMap = new HashMap<>();
            hashMap.put("10001", "小傅哥");
            hashMap.put("10002", "八杯水");
            hashMap.put("10003", "阿毛");

            return "你被代理了 " + method.getName() + ": " + hashMap.get(args[0].toString());
        };
        return (IUserDao) Proxy.newProxyInstance(Thread.currentThread().getContextClassLoader(), new Class[]{IUserDao.class}, handler);
    }

    @Override
    public Class<?> getObjectType() {
        return IUserDao.class;
    }

    @Override
    public boolean isSingleton() {
        return true;
    }

}
```

ProxyBeanFactory 是一个实现 FactoryBean 接口的代理类的名称，主要模拟了 UserDao 的原有功能，类似于 MyBatis 框架中的代理操作。

getObject 方法提供了一个 InvocationHandler 代理对象，当调用 getObject 方法时，执行 InvocationHandler 代理对象的功能。

3. 配置文件

```xml
<?xml version="1.0" encoding="UTF-8"?>
<beans>

    <bean id="userService" class="cn.bugstack.springframework.test.bean.UserService" scope="prototype">
        <property name="uId" value="10001"/>
        <property name="company" value="腾讯"/>
        <property name="location" value="深圳"/>
        <property name="userDao" ref="proxyUserDao"/>
    </bean>
```

```xml
    <bean id="proxyUserDao" class="cn.bugstack.springframework.test.bean.ProxyBeanFactory"/>
</beans>
```

在配置文件中，将 proxyUserDao 代理对象注入 userService 的 userDao 中。这里用代理类替换了 UserDao 实现类。

4. 单元测试（单例模式与原型模式）

```
@Test
public void test_prototype() {
    // 1. 初始化 BeanFactory 接口
    ClassPathXmlApplicationContext applicationContext = new ClassPathXmlApplicationContext("classpath:spring.xml");
    applicationContext.registerShutdownHook();

    // 2. 获取 Bean 对象的调用方法
    UserService userService01 = applicationContext.getBean("userService", UserService.class);
    UserService userService02 = applicationContext.getBean("userService", UserService.class);

    // 3. 配置 scope="prototype/singleton"
    System.out.println(userService01);
    System.out.println(userService02);

    // 4. 输出十六进制哈希值
    System.out.println(userService01 + " 十六进制哈希值: " + Integer.toHexString(userService01.hashCode()));
    System.out.println(ClassLayout.parseInstance(userService01).toPrintable());
}
```

在 spring.xml 配置文件中，设置了 scope="prototype"，这样每次获取的对象都是一个新对象。

判断对象是否为一个类对象的哈希值，并输出十六进制哈希值。

测试结果如下。

```
cn.bugstack.springframework.test.bean.UserService$$EnhancerByCGLIB$$4cabb984@27f674d
cn.bugstack.springframework.test.bean.UserService$$EnhancerByCGLIB$$4cabb984@2f7c7260
cn.bugstack.springframework.test.bean.UserService$$EnhancerByCGLIB$$4cabb984@27f674d
十六进制哈希值: 27f674d
cn.bugstack.springframework.test.bean.UserService$$EnhancerByCGLIB$$4cabb984 object
```

第 9 章 对象作用域和 FactoryBean

```
internals:
OFFSET  SIZE   TYPE DESCRIPTION                         VALUE
     0     4   (object header)                          01 b3 75 03 (00000001 10110011
                                                        01110101 00000011) (58045185)
     4     4   (object header)                          1b 00 00 00 (00011011 00000000
                                                        00000000 00000000) (27)
     8     4   (object header)                          9f e1 01 f8 (10011111 11100001
                                                        00000001 11111000) (-134094433)
    12     4   java.lang.String UserService.uId         (object)
    16     4   java.lang.String UserService.            (object)
                 company
    20     4   java.lang.String UserService.            (object)
                 location
    24     4   cn.bugstack.springframework.test.        (object)
                 bean.IUserDao UserService.userDao
    28     1   boolean UserService$$EnhancerByCGL       true
                 IB$$4cabb984.CGLIB$BOUND
    29     3   (alignment/padding gap)
    32     4   net.sf.cglib.proxy.NoOp UserService      (object)
                 $$EnhancerByCGLIB$$4cabb984.
                 CGLIB$CALLBACK_0
    36     4   (loss due to the next object
                 alignment)
Instance size: 40 bytes
Space losses: 3 bytes internal + 4 bytes external = 7 bytes total


Process finished with exit code 0
```

27f674d 就是十六进制哈希值。哈希值存储的结果对应的数值是倒过来的，如图 9-3 所示。

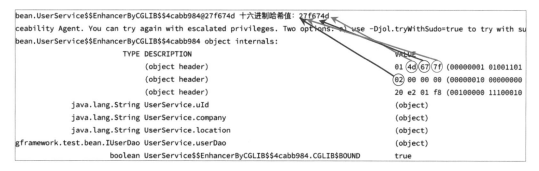

图 9-3

cabb984@27f674d 和 cabb984@2f7c7260 两个对象结尾的十六进制哈希值并不相同，所以原型模式是生效的。

5. 单元测试 (代理对象)

```
@Test
public void test_factory_bean() {
    // 1. 初始化 BeanFactory 接口
    ClassPathXmlApplicationContext applicationContext = new ClassPathXmlApplicationContext("classpath:spring.xml");
    applicationContext.registerShutdownHook();

    // 2. 调用代理方法
    UserService userService = applicationContext.getBean("userService", UserService.class);
    System.out.println(" 测试结果： " + userService.queryUserInfo());
}
```

关于 FactoryBean 的调用并没有太多不同之处，所有的不同点都已经被 spring.xml 配置文件进行了配置。也可以直接调用 spring.xml 配置文件配置的对象 cn.bugstack.springframework.test.bean.ProxyBeanFactory。

测试结果如下。

```
测试结果： 你被代理了 queryUserName: 小傅哥，腾讯，深圳

Process finished with exit code 0
```

从测试结果中可以看到，ProxyBeanFactory 代理类已经替换了 UserDao 的功能。

这个设计虽然看起来并不复杂，但是解决了所有需要和 Spring 结合的其他框架交互链接的问题。

9.5 本章总结

在 Spring 框架的整个开发过程中，前期的各个功能接口类的扩展有很多方法，虽然学习起来较难，但是后期的完善功能就没有那么难了。在深入理解相应知识后，读者会发现功能的补充都比较简单，只需要在所属领域范围内扩展相应的服务实现。

当读者仔细阅读关于 FactoryBean 的实现及测试过程的使用后，再使用 FactoryBean 开发相应的组件时，会非常清楚它是如何创建复杂的 Bean 对象，以及在什么时候进行初始化和调用的。从而快速地排查、定位和解决问题。

如果读者在学习的过程中感觉这些类、接口、实现和继承很复杂，难以理解，那么最好的方式是先画出这些类关系图，梳理实现的结构，再学习每个类的功能。

第 10 章
容器事件和事件监听器

在学习编写 Spring 框架的过程中，你可能已经体会到它在设计和实现上的复杂性，也体会到每次扩展新增的功能都会让原有的程序变得不稳定。其实，这些功能扩展的便捷性和易于维护性都来自对设计模式的巧妙使用。无论是单一职责、迪米特法则，还是依赖倒置，都是来保障程序开发过程中的高内聚和低耦合的，确保整个结构的稳定性。

本章在实现 Spring 事件的功能时，会使用观察者模式。这是一种解耦应用上下文的设计思路。当对象之间存在一对多关系时，可以使用观察者模式。它用于定义对象之间的一对多的依赖关系。当一个对象的状态发生改变时，所有依赖它的对象都会得到通知并且自动更新。

- 本章难度：★★★★☆
- 本章重点：事件和事件监听器是 Spring 框架扩展出来的非常重要的功能。它通过继承 EventObject 实现对应的容器事件，并利用统一的事件处理类，将符合用户发布类型的事件筛选出来并推送到用户监听器中，以此解耦用户业务流程的逻辑。

10.1 运用事件机制降低耦合度

在 Spring 中有一个 Event 事件功能，它可以提供事件的定义、发布及监听事件功能，并完成一些自定义的操作。例如，定义一个新用户注册的事件，当用户注册完成后，在事件监听中给用户发送一些优惠券和短信提醒，这时可以将属于基本功能的注册和对应的策略服务分开，降低系统的耦合。当需要扩展注册服务时（如添加风控策略、添加实名认证、判断用户属性等），不会影响依赖注册成功后执行的操作。

本章需要以观察者模式的方式，设计和实现 Event 的容器事件和事件监听器的功能，最终可以在现有的、已经实现的 Spring 框架中定义、监听和发布自己的事件信息。

10.2　事件观察者设计

事件的设计本身就是一种观察者模式的实现。它解决的是如何将一个对象状态改变为其他对象通知的问题，而且要考虑易用性和低耦合，保证高度的协作。

首先定义事件类、监听类、发布类，实现这些类的功能需要结合 Spring 中的 AbstractApplicationContext#refresh 方法，以便执行处理事件初始化和注册事件监听器的操作。整体设计结构如图 10-1 所示。

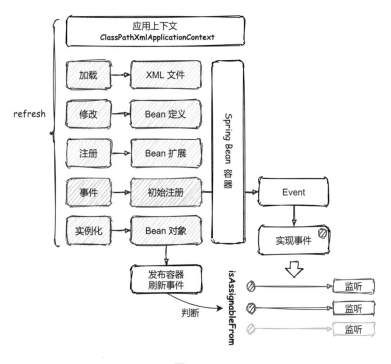

图 10-1

在整个功能的实现过程中，仍然需要在面向用户的应用上下文 AbstractApplicationContext 中添加相关事件的内容，包括初始化事件发布者、注册事件监听器、发布容器完成刷新事件。

在使用观察者模式定义事件类、监听类、发布类后，还需要实现一个广播器的功能。当接收事件推送时，对接收者感兴趣的监听事件进行分析，可以使用 isAssignableFrom 进行判断。

isAssignableFrom 的功能和 instanceof 的功能相似。isAssignableFrom 用来判断子类和父类的关系，或者接口的实现类和接口的关系。默认所有类的终极父类都是 Object 对象。如果 A.isAssignableFrom(B) 的值为 true，则证明 B 可以转换成 A，也就是说，A 可以由 B 转换而来。

10.3 事件观察者实现

1. 工程结构

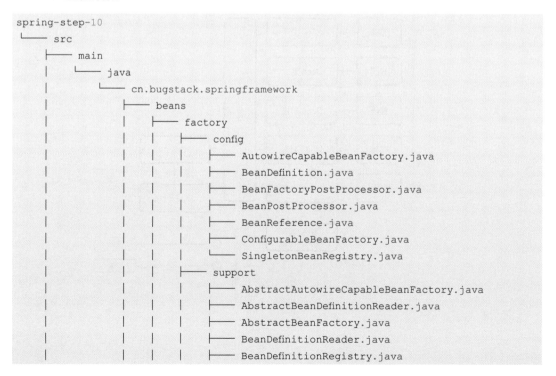

第 10 章 容器事件和事件监听器

```
│   │   │   │   ├── CglibSubclassingInstantiationStrategy.java
│   │   │   │   ├── DefaultListableBeanFactory.java
│   │   │   │   ├── DefaultSingletonBeanRegistry.java
│   │   │   │   ├── DisposableBeanAdapter.java
│   │   │   │   ├── FactoryBeanRegistrySupport.java
│   │   │   │   ├── InstantiationStrategy.java
│   │   │   │   └── SimpleInstantiationStrategy.java
│   │   │   ├── xml
│   │   │   │   └── XmlBeanDefinitionReader.java
│   │   │   ├── Aware.java
│   │   │   ├── BeanClassLoaderAware.java
│   │   │   ├── BeanFactory.java
│   │   │   ├── BeanFactoryAware.java
│   │   │   ├── BeanNameAware.java
│   │   │   ├── ConfigurableListableBeanFactory.java
│   │   │   ├── DisposableBean.java
│   │   │   ├── FactoryBean.java
│   │   │   ├── HierarchicalBeanFactory.java
│   │   │   ├── InitializingBean.java
│   │   │   └── ListableBeanFactory.java
│   │   ├── BeansException.java
│   │   ├── PropertyValue.java
│   │   └── PropertyValues.java
│   ├── context
│   │   ├── event
│   │   │   ├── AbstractApplicationEventMulticaster.java
│   │   │   ├── ApplicationContextEvent.java
│   │   │   ├── ApplicationEventMulticaster.java
│   │   │   ├── ContextClosedEvent.java
│   │   │   ├── ContextRefreshedEvent.java
│   │   │   └── SimpleApplicationEventMulticaster.java
│   │   ├── support
│   │   │   ├── AbstractApplicationContext.java
│   │   │   ├── AbstractRefreshableApplicationContext.java
│   │   │   ├── AbstractXmlApplicationContext.java
│   │   │   ├── ApplicationContextAwareProcessor.java
│   │   │   └── ClassPathXmlApplicationContext.java
│   │   ├── ApplicationContext.java
│   │   ├── ApplicationContextAware.java
│   │   ├── ApplicationEvent.java
│   │   ├── ApplicationEventPublisher.java
│   │   ├── ApplicationListener.java
```

```
|   |   |   └── ConfigurableApplicationContext.java
|   |   ├── core.io
|   |   |   ├── ClassPathResource.java
|   |   |   ├── DefaultResourceLoader.java
|   |   |   ├── FileSystemResource.java
|   |   |   ├── Resource.java
|   |   |   ├── ResourceLoader.java
|   |   |   └── UrlResource.java
|   |   └── utils
|   |       └── ClassUtils.java
└── test
    └── java
        └── cn.bugstack.springframework.test
            ├── event
            |   ├── ContextClosedEventListener.java
            |   ├── ContextRefreshedEventListener.java
            |   ├── CustomEvent.java
            |   └── CustomEventListener.java
            └── ApiTest.java
```

容器事件和事件监听器实现的类的关系如图 10-2 所示。

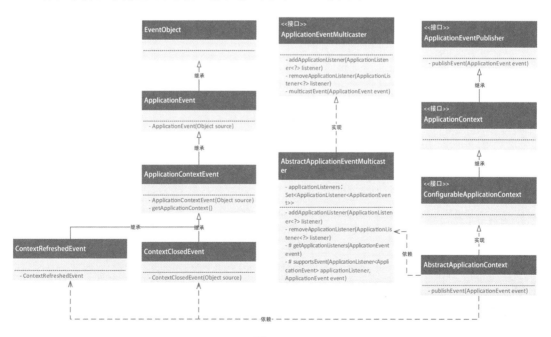

图 10-2

上述整个类的关系图围绕 Event 的容器事件定义、发布、监听功能来实现，以及使用 AbstractApplicationContext#refresh 方法对事件的相关内容进行注册和处理。

在事件的实现过程中，以扩展 Spring 的 context 包为主，在这个包中进行功能扩展。目前所有的实现内容仍然以 IOC 为主。

ApplicationContext 容器继承事件发布功能接口 ApplicationEventPublisher，并在实现类中提供事件监听功能。

ApplicationEventMulticaster 接口是注册监听器和发布事件的广播器，提供添加、移除和发布事件的功能。

发布容器关闭事件需要扩展到 AbstractApplicationContext#close 方法中，由注册到实现虚拟机的钩子。

2. 定义和实现事件

源码详见：cn.bugstack.springframework.context.ApplicationEvent。

```java
public abstract class ApplicationEvent extends EventObject {

    /**
     * Constructs a prototypical Event
     *
     * @param source the object on which the Event initially occurred
     * @throws IllegalArgumentException if source is null
     */
    public ApplicationEvent(Object source) {
        super(source);
    }

}
```

继承 java.util.EventObject 并定义具备事件功能的 ApplicationEvent 抽象类，后续所有事件的实现类都需要继承 ApplicationEvent 抽象类。

源码详见：cn.bugstack.springframework.context.event.ApplicationContextEvent。

```java
public class ApplicationContextEvent extends ApplicationEvent {

    /**
     * Constructs a prototypical Event
     *
```

```java
     * @param source the object on which the Event initially occurred
     * @throws IllegalArgumentException if source is null
     */
    public ApplicationContextEvent(Object source) {
        super(source);
    }

    /**
     * Get the <code>ApplicationContext</code> that the event was raised for
     */
    public final ApplicationContext getApplicationContext() {
        return (ApplicationContext) getSource();
    }

}
```

源码详见：cn.bugstack.springframework.context.event.ContextClosedEvent。

```java
public class ContextClosedEvent extends ApplicationContextEvent{

    /**
     * Constructs a prototypical Event.
     *
     * @param source the object on which the Event initially occurred
     * @throws IllegalArgumentException if source is null
     */
    public ContextClosedEvent(Object source) {
        super(source);
    }

}
```

源码详见：cn.bugstack.springframework.context.event.ContextRefreshedEvent。

```java
public class ContextRefreshedEvent extends ApplicationContextEvent{
    /**
     * Constructs a prototypical Event.
     *
     * @param source the object on which the Event initially occurred
     * @throws IllegalArgumentException if source is null
     */
    public ContextRefreshedEvent(Object source) {
        super(source);
    }

}
```

ApplicationContextEvent 定义事件的抽象类，包括关闭、刷新及用户自己实现的事件，都需要继承这个类。

ContextClosedEvent、ContextRefreshedEvent 分别是 Spring 框架实现的两个事件类，可用于监听关闭和刷新。

3. 事件广播器

源码详见：cn.bugstack.springframework.context.event.ApplicationEventMulticaster。

```java
public interface ApplicationEventMulticaster {

    /**
     * Add a listener to be notified of all events
     * @param listener the listener to add
     */
    void addApplicationListener(ApplicationListener<?> listener);

    /**
     * Remove a listener from the notification list
     * @param listener the listener to remove
     */
    void removeApplicationListener(ApplicationListener<?> listener);

    /**
     * Multicast the given application event to appropriate listeners
     * @param event the event to multicast
     */
    void multicastEvent(ApplicationEvent event);

}
```

在事件广播器中，定义了添加监听和删除监听的方法及一个广播事件的方法 multicastEvent，最终通过 multicastEvent 方法来推送时间消息，决定由谁接收事件。

源码详见：cn.bugstack.springframework.context.event.AbstractApplicationEventMulticaster。

```java
public abstract class AbstractApplicationEventMulticaster implements
ApplicationEventMulticaster, BeanFactoryAware {

    public final Set<ApplicationListener<ApplicationEvent>> applicationListeners =
new LinkedHashSet<>();
```

```java
    private BeanFactory beanFactory;

    @Override
    public void addApplicationListener(ApplicationListener<?> listener) {
        applicationListeners.add((ApplicationListener<ApplicationEvent>) listener);
    }

    @Override
    public void removeApplicationListener(ApplicationListener<?> listener) {
        applicationListeners.remove(listener);
    }

    @Override
    public final void setBeanFactory(BeanFactory beanFactory) {
        this.beanFactory = beanFactory;
    }

    protected Collection<ApplicationListener> getApplicationListeners(ApplicationEvent event) {
        LinkedList<ApplicationListener> allListeners = new LinkedList<ApplicationListener>();
        for (ApplicationListener<ApplicationEvent> listener : applicationListeners) {
            if (supportsEvent(listener, event)) allListeners.add(listener);
        }
        return allListeners;
    }

    /**
     * 监听器是否对该事件感兴趣
     */
    protected boolean supportsEvent(ApplicationListener<ApplicationEvent> applicationListener, ApplicationEvent event) {
        Class<? extends ApplicationListener> listenerClass = applicationListener.getClass();

        // 按照 CglibSubclassingInstantiationStrategy 和 SimpleInstantiationStrategy 不
        // 同的实例化类型，需要判断后获取目标 class
        Class<?> targetClass = ClassUtils.isCglibProxyClass(listenerClass) ? listenerClass.getSuperclass() : listenerClass;
        Type genericInterface = targetClass.getGenericInterfaces()[0];

        Type actualTypeArgument = ((ParameterizedType) genericInterface).
```

```
getActualTypeArguments()[0];
    String className = actualTypeArgument.getTypeName();
    Class<?> eventClassName;
    try {
        eventClassName = Class.forName(className);
    } catch (ClassNotFoundException e) {
        throw new BeansException("wrong event class name: " + className);
    }
    // 判断此 eventClassName 对象表示的类或接口与指定的 event.getClass 参数所表示的类或接
    // 口是否相同，或者是否是其超类或超接口
    // isAssignableFrom 用来判断子类和父类的关系，或者接口的实现类和接口的关系
    // 默认所有类的终极父类都是 Object
    // 如果 A.isAssignableFrom(B) 的值为 true，则证明 B 可以转换成 A，也就是说，A 可以由 B 转换而来
    return eventClassName.isAssignableFrom(event.getClass());
    }
}
```

AbstractApplicationEventMulticaster 用于对事件广播器的公用方法进行提取，这个类可以实现一些基本功能，避免所有直接实现接口处理细节。这个类除了用于处理 addApplicationListener、removeApplicationListener 这样的通用方法，还用于处理 getApplicationListeners 方法和 supportsEvent 方法。

getApplicationListeners 方法的作用是选取符合广播事件中的监听处理器，具体的过滤操作在 supportsEvent 方法中实现。

supportsEvent 方法主要包括 Cglib 和 Simple 不同实例化的类型，通过判断后获取目标 Class。Cglib 代理类需要获取父类的 Class，普通实例化的类则不需要获取父类的 Class。通过提取接口和对应的 ParameterizedType 类和 eventClassName 类，确认两者是否为子类和父类的关系，以此将此事件交给符合的类处理。具体可以参考代码中的注释。

图 10-3 所示为 supportsEvent 方法的运行图。

在代码调试中可以看到，最终由 isAssignableFrom 判断 eventClassName 和 event.getClass 的值为 true。对于 CglibSubclassingInstantiationStrategy、SimpleInstantiationStrategy 来说，可以尝试在 AbstractApplicationContext 类中更换验证。

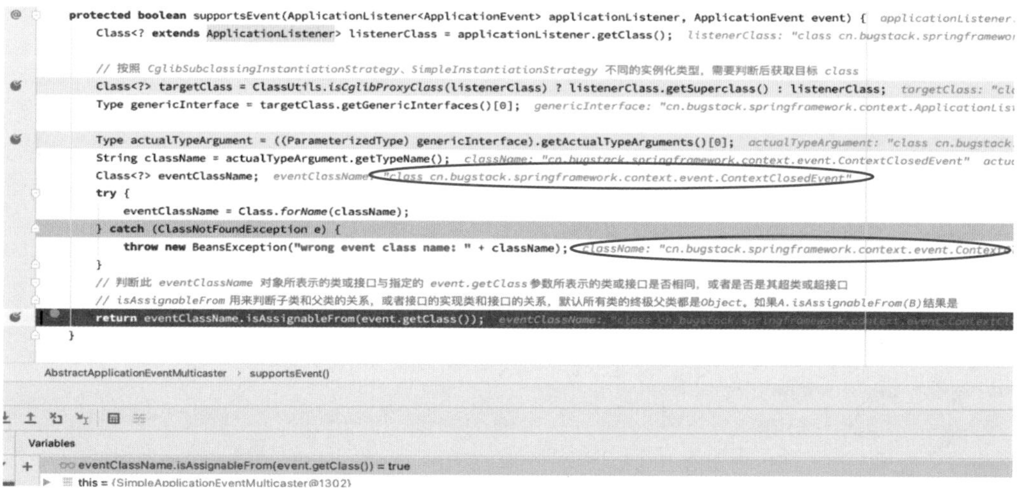

```
 protected boolean supportsEvent(ApplicationListener<ApplicationEvent> applicationListener, ApplicationEvent event) { applicationListener...
     Class<? extends ApplicationListener> listenerClass = applicationListener.getClass(); listenerClass: "class cn.bugstack.springframewor
     // 按照 CglibSubclassingInstantiationStrategy、SimpleInstantiationStrategy 不同的实例化类型，需要判断后获取目标 class
     Class<?> targetClass = ClassUtils.isCglibProxyClass(listenerClass) ? listenerClass.getSuperclass() : listenerClass; targetClass: "cla
     Type genericInterface = targetClass.getGenericInterfaces()[0]; genericInterface: "cn.bugstack.springframework.context.ApplicationLis

     Type actualTypeArgument = ((ParameterizedType) genericInterface).getActualTypeArguments()[0]; actualTypeArgument: "class cn.bugstack
     String className = actualTypeArgument.getTypeName(); className: "cn.bugstack.springframework.context.event.ContextClosedEvent" actu
     Class<?> eventClassName;   eventClassName = "class cn.bugstack.springframework.context.event.ContextClosedEvent"
     try {
         eventClassName = Class.forName(className);
     } catch (ClassNotFoundException e) {
         throw new BeansException("wrong event class name: " + className); className: "cn.bugstack.springframework.context.event.Contex
     }
     // 判断此 eventClassName 对象所表示的类或接口与指定的 event.getClass 参数所表示的类或接口是否相同，或者是否是其超类或超接口
     // isAssignableFrom 用来判断子类和父类的关系，或者接口的实现类和接口的关系，默认所有类的终极父类都是Object。如果A.isAssignableFrom(B)结果是
     return eventClassName.isAssignableFrom(event.getClass()); eventClassName: "class cn.bugstack.springframework.context.event.ContextC
 }

AbstractApplicationEventMulticaster > supportsEvent()

Variables
    eventClassName.isAssignableFrom(event.getClass()) = true
    this = {SimpleApplicationEventMulticaster@1302}
```

图 10-3

4. 事件发布者的定义和实现

源码详见：cn.bugstack.springframework.context.ApplicationEventPublisher。

```
public interface ApplicationEventPublisher {

    /**
     * Notify all listeners registered with this application of an application
     * event. Events may be framework events (such as RequestHandledEvent)
     * or application-specific events.
     * @param event the event to publish
     */
    void publishEvent(ApplicationEvent event);

}
```

ApplicationEventPublisher 是事件的发布接口，所有的事件都需要从这个接口发布出去。

源码详见：cn.bugstack.springframework.context.support.AbstractApplicationContext。

```
public abstract class AbstractApplicationContext extends DefaultResourceLoader
implements ConfigurableApplicationContext {

    public static final String APPLICATION_EVENT_MULTICASTER_BEAN_NAME = "applicationEventMulticaster";
```

```java
    private ApplicationEventMulticaster applicationEventMulticaster;

    @Override
    public void refresh() throws BeansException {

        // 1. 初始化事件发布者
        initApplicationEventMulticaster();

        // 2. 注册事件监听器
        registerListeners();

        // 3. 发布容器完成刷新事件
        finishRefresh();
    }

    private void initApplicationEventMulticaster() {
        ConfigurableListableBeanFactory beanFactory = getBeanFactory();
        applicationEventMulticaster = new SimpleApplicationEventMulticaster(beanFactory);
        beanFactory.registerSingleton(APPLICATION_EVENT_MULTICASTER_BEAN_NAME, applicationEventMulticaster);
    }

    private void registerListeners() {
        Collection<ApplicationListener> applicationListeners = getBeansOfType(ApplicationListener.class).values();
        for (ApplicationListener listener : applicationListeners) {
            applicationEventMulticaster.addApplicationListener(listener);
        }
    }

    private void finishRefresh() {
        publishEvent(new ContextRefreshedEvent(this));
    }

    @Override
    public void publishEvent(ApplicationEvent event) {
        applicationEventMulticaster.multicastEvent(event);
    }

    @Override
    public void close() {
        // 容器关闭事件
```

```
        publishEvent(new ContextClosedEvent(this));

        // 执行销毁单例Bean对象的销毁方法
        getBeanFactory().destroySingletons();
    }
}
```

在抽象应用上下文 AbstractApplicationContext#refresh 中，新增了初始化事件发布者、注册事件监听器、发布容器完成刷新事件，用于处理事件操作。

初始化事件发布者（initApplicationEventMulticaster）主要用于实例化一个 SimpleApplicationEventMulticaster 事件广播器。

注册事件监听器（registerListeners）通过 getBeansOfType 方法获取所有从 spring.xml 配置文件中加载的事件配置 Bean 对象。

发布容器完成刷新事件（finishRefresh）发布了第一个服务器启动完成后的事件，这个事件通过 publishEvent 方法发布出去，也就是调用了 applicationEventMulticaster.multicastEvent(event) 方法。

close 方法新增了一个容器关闭事件 publishEvent(new ContextClosedEvent(this))。

10.4 事件使用测试

1. 创建一个事件和监听器

源码详见：cn.bugstack.springframework.test.event.CustomEvent。

```
public class CustomEvent extends ApplicationContextEvent {

    private Long id;
    private String message;

    /**
     * Constructs a prototypical Event
     *
     * @param source the object on which the Event initially occurred
     * @throws IllegalArgumentException if source is null
     */
```

```java
public CustomEvent(Object source, Long id, String message) {
    super(source);
    this.id = id;
    this.message = message;
}

// …get/set
}
```

创建一个自定义事件，在事件类的构造函数中可以添加想要的属性信息。由于 CustomEvent 这个事件类最终会被完整地输入监听器中，所以添加的属性都会被获取。

源码详见：cn.bugstack.springframework.test.event.CustomEventListener。

```java
public class CustomEventListener implements ApplicationListener<CustomEvent> {

    @Override
    public void onApplicationEvent(CustomEvent event) {
        System.out.println("收到：" + event.getSource() + " 消息;时间：" + new Date());
        System.out.println(" 消息：" + event.getId() + ":" + event.getMessage());
    }

}
```

这是一个用于监听 CustomEvent 事件的监听器，可在用户注册后用于发送优惠券和短信通知等。ContextRefreshedEventListener implements ApplicationListener 监听器和 ContextClosedEventListener implements ApplicationListener 监听器不再逐一演示，读者可以参考源码。

2. 配置文件

```xml
<?xml version="1.0" encoding="UTF-8"?>
<beans>

    <bean class="cn.bugstack.springframework.test.event.ContextRefreshedEventListener"/>

    <bean class="cn.bugstack.springframework.test.event.CustomEventListener"/>

    <bean class="cn.bugstack.springframework.test.event.ContextClosedEventListener"/>

</beans>
```

在 spring.xml 配置文件中配置了 3 个事件监听器——监听刷新、监听自定义事件、监听关闭事件。

3. 单元测试

```
public class ApiTest {

    @Test
    public void test_event() {
        ClassPathXmlApplicationContext applicationContext = new 
ClassPathXmlApplicationContext("classpath:spring.xml");
        applicationContext.publishEvent(new CustomEvent(applicationContext, 
1019129009086763L, "成功了！"));

        applicationContext.registerShutdownHook();
    }

}
```

用户通过使用 applicationContext 新增的发布事件接口方法，发布一个自定义事件 CustomEvent，并传递相应的参数信息。

测试结果如下。

```
刷新事件: cn.bugstack.springframework.test.event.
ContextRefreshedEventListener$$EnhancerByCGLIB$$440a36f5
收到: cn.bugstack.springframework.context.support.ClassPathXmlApplicationContext@71c7db30
消息；时间: 22:32:50
消息: 1019129009086763: 成功了！
关闭事件: cn.bugstack.springframework.test.event.
ContextClosedEventListener$$EnhancerByCGLIB$$f4d4b18d

Process finished with exit code 0
```

从测试结果中可以看到，自定义的事件及监听系统的事件信息，都可以在控制台中完整地输出了。读者也可以尝试增加一些其他事件，并调试代码至学习观察者模式。

10.5 本章总结

在学习 Spring 框架的过程中，读者可以看到很多设计模式，如 BeanFactory、FactoryBean、策略模式访问资源与观察者模式。所以要更加注重设计模式的运用，这既有助于读懂代码，也是学习的重点。

本章中观察者模式的实现过程主要包括事件的定义、监听和发布，发布完成后根据匹配策略，监听器收到属于自己的事件内容，并进行相应的处理。观察者模式在日常应用中也被经常使用，在结合 Spring 框架之后，读者除了可以学习到设计模式，还可以学习到如何把相应观察者的实现和应用上下文结合起来。

所有在 Spring 框架中学习到的技术、设计、思路都可以和实际的业务开发结合起来，这些看似比较多的代码模块，其实是按照各自的职责一点一点地扩充进去的。在学习过程中，读者可以先尝试完成这些框架的功能，再逐步通过调试的方式与 Spring 源码进行对照参考，从而掌握这些设计和编码能力。

第 11 章
基于 JDK、Cglib 实现 AOP 切面

Spring 包括一纵一横两大思想——IOC 和 AOP。我们日常使用的功能都离不开它们的支持。在前文介绍的功能开发中，已经基本实现了一个完整的 IOC 功能结构。那么，有了 IOC，为什么还要添加 AOP 呢？

当为多个不存在继承关系的功能对象类提供一个统一的行为操作时，会有大量的重复代码开发或者反复调用。例如，方法监控、接口管理、日志打印等，都需要在代码块中引入相应的非业务逻辑的同类功能处理，程序代码将越来越难以维护。所以，除了 IOC 的纵向处理，也需要 AOP 横向切面进行统一的共性逻辑处理，简化程序的开发过程。

- 本章难度：★★★★☆
- 本章重点：基于类的代理和方法拦截匹配器，为目标对象提供 AOP 切面处理的统一框架结构。在学习中注意先厘清类的代理知识，再学习 AOP 切面的设计思想和技术实现。

11.1 动态代理

下面介绍关于 AOP（Aspect-oriented programming，面向切面编程）内容的开发。AOP 通过预编译的方式和运行期间动态代理，实现程序功能的统一维护。其实，AOP 是 OOP 的延续，是 Spring 框架中一个非常重要的内容。使用 AOP 可以对业务逻辑的各个部分进行隔离，从而降低各模块之间业务逻辑的耦合度，提高代码的可复用性，也能提高程序的开发效率。

AOP 的核心技术主要是动态代理的使用，就像对于一个接口的实现类，可以使用代理类的方式将其替换，处理需要的逻辑。例如：

```
@Test
public void test_proxy_class() {
    IUserService userService = (IUserService) Proxy.newProxyInstance(Thread.
currentThread().getContextClassLoader(), new Class[]{IUserService.class}, (proxy,
method, args) -> "你被代理了！ ");
    String result = userService.queryUserInfo();
    System.out.println("测试结果: " + result);
}
```

实现了代理类后，接下来就需要考虑怎么给方法做代理，而不是代理整个类。如果想要代理所有类的方法，则可以制作一个方法拦截器，给所有被代理的方法添加一些自定义处理，如输出日志、记录耗时、监控异常等。

11.2　AOP 切面设计

在将 AOP 整个切面设计融合到 Spring 框架中之前，需要解决两个问题：如何给符合规则的方法做代理？以及做完代理方法的实例后，如何把类的职责拆分出来？这两个功能都是以切面的思想进行设计和开发的。如果不清楚什么是 AOP，则可以把切面理解为用刀切韭菜，一根一根切有点慢。如果用手（代理）将韭菜捏成一把，用不同的拦截操作（用菜刀切）来处理，就会达到事半功倍的效果。在程序中也一样，只不过韭菜变成了方法，菜刀变成了拦截方法。整体的设计结构如图 11-1 所示。

就像使用 Spring 的 AOP 一样，只处理一些需要被拦截的方法。在执行完拦截方法后，执行方法的扩展操作。首先实现一个可以代理方法的 Proxy。代理方法主要是使用方法拦截器类处理方法的调用 MethodInterceptor#invoke，而不是直接使用 invoke 方法中的入参信息 Method method 执行 method.invoke(targetObj, args) 操作。除了要实现以上核心功能，还需要使用 org.aspectj.weaver.tools.PointcutParser 处理拦截表达式 "execution(* cn.bugstack.springframework.test.bean.IUserService.*(..))"。有了方法拦截器和处理器，就可以设计出一个 AOP 的雏形。

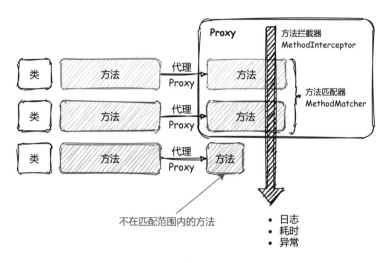

图 11-1

11.3 AOP 切面实现

1. 工程结构

```
spring-step-11
└── src
    └── main
        └── java
            └── cn.bugstack.springframework
                ├── aop
                │   ├── aspectj
                │   │   └── AspectJExpressionPointcut.java
                │   ├── framework
                │   │   ├── AopProxy.java
                │   │   ├── Cglib2AopProxy.java
                │   │   ├── JdkDynamicAopProxy.java
                │   │   └── ReflectiveMethodInvocation.java
                │   ├── AdvisedSupport.java
                │   ├── ClassFilter.java
                │   ├── MethodMatcher.java
                │   ├── Pointcut.java
                │   └── TargetSource.java
                └── beans
```

第 11 章 基于 JDK、Cglib 实现 AOP 切面

```
|   |   ├── factory
|   |   |   ├── config
|   |   |   |   ├── AutowireCapableBeanFactory.java
|   |   |   |   ├── BeanDefinition.java
|   |   |   |   ├── BeanFactoryPostProcessor.java
|   |   |   |   ├── BeanPostProcessor.java
|   |   |   |   ├── BeanReference.java
|   |   |   |   ├── ConfigurableBeanFactory.java
|   |   |   |   └── SingletonBeanRegistry.java
|   |   |   ├── support
|   |   |   |   ├── AbstractAutowireCapableBeanFactory.java
|   |   |   |   ├── AbstractBeanDefinitionReader.java
|   |   |   |   ├── AbstractBeanFactory.java
|   |   |   |   ├── BeanDefinitionReader.java
|   |   |   |   ├── BeanDefinitionRegistry.java
|   |   |   |   ├── CglibSubclassingInstantiationStrategy.java
|   |   |   |   ├── DefaultListableBeanFactory.java
|   |   |   |   ├── DefaultSingletonBeanRegistry.java
|   |   |   |   ├── DisposableBeanAdapter.java
|   |   |   |   ├── FactoryBeanRegistrySupport.java
|   |   |   |   ├── InstantiationStrategy.java
|   |   |   |   └── SimpleInstantiationStrategy.java
|   |   |   ├── xml
|   |   |   |   └── XmlBeanDefinitionReader.java
|   |   |   ├── Aware.java
|   |   |   ├── BeanClassLoaderAware.java
|   |   |   ├── BeanFactory.java
|   |   |   ├── BeanFactoryAware.java
|   |   |   ├── BeanNameAware.java
|   |   |   ├── ConfigurableListableBeanFactory.java
|   |   |   ├── DisposableBean.java
|   |   |   ├── FactoryBean.java
|   |   |   ├── HierarchicalBeanFactory.java
|   |   |   ├── InitializingBean.java
|   |   |   └── ListableBeanFactory.java
|   |   ├── BeansException.java
|   |   ├── PropertyValue.java
|   |   └── PropertyValues.java
|   ├── context
|   |   ├── event
|   |   |   ├── AbstractApplicationEventMulticaster.java
|   |   |   ├── ApplicationContextEvent.java
|   |   |   ├── ApplicationEventMulticaster.java
|   |   |   ├── ContextClosedEvent.java
```

```
|   |   |   |   ├── ContextRefreshedEvent.java
|   |   |   |   └── SimpleApplicationEventMulticaster.java
|   |   |   ├── support
|   |   |   |   ├── AbstractApplicationContext.java
|   |   |   |   ├── AbstractRefreshableApplicationContext.java
|   |   |   |   ├── AbstractXmlApplicationContext.java
|   |   |   |   ├── ApplicationContextAwareProcessor.java
|   |   |   |   └── ClassPathXmlApplicationContext.java
|   |   |   ├── ApplicationContext.java
|   |   |   ├── ApplicationContextAware.java
|   |   |   ├── ApplicationEvent.java
|   |   |   ├── ApplicationEventPublisher.java
|   |   |   ├── ApplicationListener.java
|   |   |   └── ConfigurableApplicationContext.java
|   |   ├── core.io
|   |   |   ├── ClassPathResource.java
|   |   |   ├── DefaultResourceLoader.java
|   |   |   ├── FileSystemResource.java
|   |   |   ├── Resource.java
|   |   |   ├── ResourceLoader.java
|   |   |   └── UrlResource.java
|   |   └── utils
|   |       └── ClassUtils.java
└── test
    └── java
        └── cn.bugstack.springframework.test
            ├── bean
            |   ├── IUserService.java
            |   ├── UserService.java
            |   └── UserServiceInterceptor.java
            └── ApiTest.java
```

　　AOP 切点的表达式和使用方法，以及基于 JDK 和 Cglib 的动态代理类的关系如图 11-2 所示。

　　整个类关系图就是 AOP 实现的核心逻辑，上面部分是关于方法的匹配实现，下面部分从 AopProxy 接口开始是关于方法的代理操作。AspectJExpressionPointcut 的核心功能主要依赖于 aspectj 组件，实现了 Pointcut 接口、ClassFilter 接口、MethodMatcher 接口，专门用于处理类和方法的匹配过滤操作。AopProxy 是代理的抽象对象，它主要基于 JDK 代理和 Cglib 代理实现。在前文介绍的关于对象的实例化 CglibSubclassingInstantiationStrategy 中，也使用过 Cglib 提供的功能。

第 11 章 基于 JDK、Cglib 实现 AOP 切面

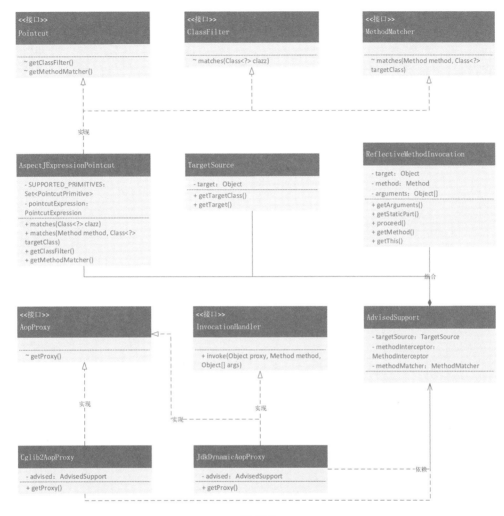

图 11-2

2. 代理方法实例

在实现 AOP 的核心功能之前，我们先通过一个代理方法的实例了解其全貌，以便更好地理解后续拆解各个方法，以及设计具有解耦功能的 AOP 实现过程。

（1）单元测试。

```
@Test
public void test_proxy_method() {
    // 目标对象（可以替换成任何的目标对象）
```

155

```java
        Object targetObj = new UserService();
        // AOP 代理
        IUserService proxy = (IUserService) Proxy.newProxyInstance(Thread.currentThread().
getContextClassLoader(), targetObj.getClass().getInterfaces(), new InvocationHandler() {
            // 方法匹配器
            MethodMatcher methodMatcher = new AspectJExpressionPointcut("execution(* 
cn.bugstack.springframework.test.bean.IUserService.*(..))");
            @Override
            public Object invoke(Object proxy, Method method, Object[] args) throws 
Throwable {
                if (methodMatcher.matches(method, targetObj.getClass())) {
                    // 方法拦截器
                    MethodInterceptor methodInterceptor = invocation -> {
                        long start = System.currentTimeMillis();
                        try {
                            return invocation.proceed();
                        } finally {
                            System.out.println("监控 - Begin By AOP");
                            System.out.println("方法名称: " + invocation.getMethod().getName());
                            System.out.println("方法耗时: " + (System.currentTimeMillis() 
- start) + "ms");
                            System.out.println("监控 - End\r\n");
                        }
                    };
                    // 反射调用
                    return methodInterceptor.invoke(new ReflectiveMethodInvocation
(targetObj, method, args));
                }
                return method.invoke(targetObj, args);
            }
        });
        String result = proxy.queryUserInfo();
        System.out.println("测试结果: " + result);
}
```

整个实例的目标是将一个 UserService 作为目标对象，对类中的所有方法进行拦截，并添加监控信息，输出打印。实例中有代理对象 Proxy.newProxyInstance、方法匹配 MethodMatcher、反射调用 invoke(Object proxy, Method method, Object[] args)，以及用户自己实现拦截方法后的操作。这样就与使用 AOP 非常类似了，只不过在使用 AOP 时，框架已经提供了更好的功能，这里是把所有的核心过程重新展示出来。

（2）测试结果。

```
监控 - Begin By AOP
方法名称：queryUserInfo
方法耗时：86ms
监控 - End

测试结果：小傅哥，100001，深圳

Process finished with exit code 0
```

从测试结果中可以看到，AOP 代理已经对 IUserService 接口的代理对象进行了拦截监控，其实后面章节中实现的 AOP 就是现在测试的结果，只是需要将这部分测试的实例拆解为扩展性更好的各个模块。

图 11-3 所示为拆解实例。

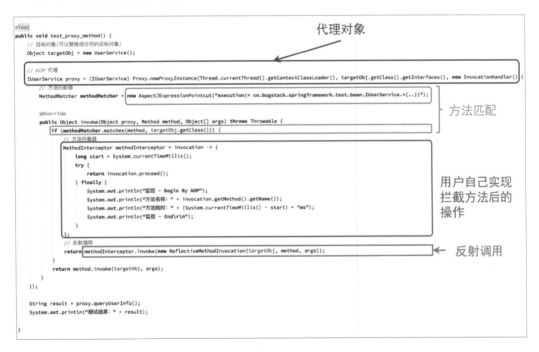

图 11-3

拆解过程如图 11-3 所示，下面将代理对象拆解出来。代理对象既可以是 JDK 的实现，也可以是 Cglib 的处理。方法匹配器已经是一个单独的实现类，需要将传入的目标对象、方法匹配、拦截方法进行统一包装，方便外部调用时进行入参透传。

ReflectiveMethodInvocation 目前已经实现 MethodInvocation 接口的一个包装后的类，参数信息包括调用的对象、方法和入参。

3. 切点表达式

（1）定义接口。

源码详见：cn.bugstack.springframework.aop.Pointcut。

```java
public interface Pointcut {

    /**
     * Return the ClassFilter for this pointcut.
     * @return the ClassFilter (never <code>null</code>)
     */
    ClassFilter getClassFilter();

    /**
     * Return the MethodMatcher for this pointcut.
     * @return the MethodMatcher (never <code>null</code>)
     */
    MethodMatcher getMethodMatcher();

}
```

切入点接口 Pointcut 用于获取 ClassFilter 和 MethodMatcher 两个类，它们都是切点表达式提供的内容。

源码详见：cn.bugstack.springframework.aop.ClassFilter。

```java
public interface ClassFilter {

    /**
     * Should the pointcut apply to the given interface or target class?
     * @param clazz the candidate target class
     * @return whether the advice should apply to the given target class
     */
    boolean matches(Class<?> clazz);

}
```

ClassFilter 接口定义匹配类，用于帮助切点找到给定的接口和目标类。

源码详见：cn.bugstack.springframework.aop.MethodMatcher。

```java
public interface MethodMatcher {
```

```
/**
 * Perform static checking whether the given method matches. If this
 * @return whether or not this method matches statically
 */
boolean matches(Method method, Class<?> targetClass);
}
```

方法匹配用于找到表达式范围内匹配的目标类和方法，在上面的实例中体现为 methodMatcher.matches(method, targetObj.getClass())。

（2）实现切点表达式类。

```
public class AspectJExpressionPointcut implements Pointcut, ClassFilter, MethodMatcher {

    private static final Set<PointcutPrimitive> SUPPORTED_PRIMITIVES = new HashSet<PointcutPrimitive>();

    static {
        SUPPORTED_PRIMITIVES.add(PointcutPrimitive.EXECUTION);
    }

    private final PointcutExpression pointcutExpression;

    public AspectJExpressionPointcut(String expression) {
        PointcutParser pointcutParser = PointcutParser.getPointcutParserSupportingSpecifiedPrimitivesAndUsingSpecifiedClassLoaderForResolution(SUPPORTED_PRIMITIVES, this.getClass().getClassLoader());
        pointcutExpression = pointcutParser.parsePointcutExpression(expression);
    }

    @Override
    public boolean matches(Class<?> clazz) {
        return pointcutExpression.couldMatchJoinPointsInType(clazz);
    }

    @Override
    public boolean matches(Method method, Class<?> targetClass) {
        return pointcutExpression.matchesMethodExecution(method).alwaysMatches();
    }

    @Override
    public ClassFilter getClassFilter() {
```

```
        return this;
    }

    @Override
    public MethodMatcher getMethodMatcher() {
        return this;
    }
}
```

切点表达式实现了 Pointcut 接口、ClassFilter 接口、MethodMatcher 接口的定义方法，同时主要使用了 aspectj 包提供的表达式校验方法。匹配 matches 包括 pointcutExpression.couldMatchJoinPointsInType(clazz) 方法、pointcutExpression.matchesMethodExecution (method).alwaysMatches 方法，这部分内容可以单独进行匹配验证。

（3）匹配验证。

```
@Test
public void test_aop() throws NoSuchMethodException {
    AspectJExpressionPointcut pointcut = new AspectJExpressionPointcut("execution(* cn.bugstack.springframework.test.bean.UserService.*(..))");
    Class<UserService> clazz = UserService.class;
    Method method = clazz.getDeclaredMethod("queryUserInfo");

    System.out.println(pointcut.matches(clazz));
    System.out.println(pointcut.matches(method, clazz));
}
```

这里单独使用一个匹配方法来验证，查看拦截的方法与对应的对象是否匹配。

4．包装切面通知信息

源码详见：cn.bugstack.springframework.aop.AdvisedSupport。

```
public class AdvisedSupport {

    // 被代理的目标对象
    private TargetSource targetSource;
    // 方法拦截器
    private MethodInterceptor methodInterceptor;
    // 方法匹配器（检查目标方法是否符合通知条件）
    private MethodMatcher methodMatcher;
```

```
    // …get/set
}
```

AdvisedSupport 将代理、拦截、匹配的各项属性包装到一个类中,方便在 Proxy 实现类中使用。这与在业务开发中包装入参是一个道理。TargetSource 是一个目标对象,在目标对象类中提供 Object 入参属性,以及获取目标类 TargetClass 的信息。Method Interceptor 是一个具体拦截方法的实现类,由用户实现 MethodInterceptor#invoke 方法并进行具体的处理。本节的实例是对方法进行监控处理。MethodMatcher 是一个匹配方法的操作,这个对象由 AspectJExpressionPointcut 提供服务。

5. 代理抽象实现(JDK&Cglib)

源码详见:cn.bugstack.springframework.aop.framework。

```
public interface AopProxy {

    Object getProxy();

}
```

定义一个标准接口,用于获取代理类。因为具体实现代理的方式既可以是 JDK 方式,也可以是 Cglib 方式,所以定义接口会更加方便管理实现类。

源码详见:cn.bugstack.springframework.aop.framework.JdkDynamicAopProxy。

```
public class JdkDynamicAopProxy implements AopProxy, InvocationHandler {

    private final AdvisedSupport advised;

    public JdkDynamicAopProxy(AdvisedSupport advised) {
        this.advised = advised;
    }

    @Override
    public Object getProxy() {
        return Proxy.newProxyInstance(Thread.currentThread().getContextClassLoader(), advised.getTargetSource().getTargetClass(), this);
    }

    @Override
    public Object invoke(Object proxy, Method method, Object[] args) throws Throwable {
        if (advised.getMethodMatcher().matches(method, advised.getTargetSource().getTarget().getClass())) {
            MethodInterceptor methodInterceptor = advised.getMethodInterceptor();
```

```
            return methodInterceptor.invoke(new ReflectiveMethodInvocation(advised.
getTargetSource().getTarget(), method, args));
        }
        return method.invoke(advised.getTargetSource().getTarget(), args);
    }

}
```

基于 JDK 实现的代理类需要实现 AopProxy 接口、InvocationHandler 接口，这样就可以将代理对象 getProxy 和反射调用方法 invoke 分开进行处理。getProxy 方法是代理一个对象的操作，需要提供入参 ClassLoader、AdvisedSupport 及 this 类，其中 this 类提供了 invoke 方法。invoke 方法在处理完匹配的方法后，利用用户提供的方法实现拦截，并进行反射调用 methodInterceptor.invoke。这里的 ReflectiveMethodInvocation 是一个入参的包装信息，提供了入参对象——目标对象、方法。

源码详见：cn.bugstack.springframework.aop.framework.Cglib2AopProxy。

```
public class Cglib2AopProxy implements AopProxy {

    private final AdvisedSupport advised;

    public Cglib2AopProxy(AdvisedSupport advised) {
        this.advised = advised;
    }

    @Override
    public Object getProxy() {
        Enhancer enhancer = new Enhancer();
        enhancer.setSuperclass(advised.getTargetSource().getTarget().getClass());
        enhancer.setInterfaces(advised.getTargetSource().getTargetClass());
        enhancer.setCallback(new DynamicAdvisedInterceptor(advised));
        return enhancer.create();
    }

    private static class DynamicAdvisedInterceptor implements MethodInterceptor {

        @Override
        public Object intercept(Object proxy, Method method, Object[] objects, MethodProxy
methodProxy) throws Throwable {
            CglibMethodInvocation methodInvocation = new CglibMethodInvocation
(advised.getTargetSource().getTarget(), method, objects, methodProxy);
            if (advised.getMethodMatcher().matches(method, advised.getTargetSource().
getTarget().getClass())) {
```

```java
            return advised.getMethodInterceptor().invoke(methodInvocation);
        }
        return methodInvocation.proceed();
    }
}

private static class CglibMethodInvocation extends ReflectiveMethodInvocation {

    @Override
    public Object proceed() throws Throwable {
        return this.methodProxy.invoke(this.target, this.arguments);
    }

}

}
```

由于基于 Cglib 使用 Enhancer 代理的类可以在运行期间为接口使用底层 ASM 字节码，来增强技术处理对象生成代理对象的功能，因此被代理类不需要实现任何接口。这里可以看到 DynamicAdvisedInterceptor#intercept 在匹配方法后进行了相应的反射操作。

11.4 AOP 切面测试

1. 事先准备

```java
public class UserService implements IUserService {

    public String queryUserInfo() {
        try {
            Thread.sleep(new Random(1).nextInt(100));
        } catch (InterruptedException e) {
            e.printStackTrace();
        }
        return "小傅哥，100001，深圳";
    }

    public String register(String userName) {
        try {
            Thread.sleep(new Random(1).nextInt(100));
        } catch (InterruptedException e) {
            e.printStackTrace();
```

```
        }
        return "注册用户: " + userName + " success! ";
    }
}
```

UserService 提供了两个不同的方法，可以在测试中增加新的类。后面的测试过程会为这两个方法添加拦截处理，并输出方法耗时。

2. 自定义拦截方法

```
public class UserService implements IUserService {

    public String queryUserInfo() {
        try {
            Thread.sleep(new Random(1).nextInt(100));
        } catch (InterruptedException e) {
            e.printStackTrace();
        }
        return "小傅哥, 100001, 深圳";
    }

    public String register(String userName) {
        try {
            Thread.sleep(new Random(1).nextInt(100));
        } catch (InterruptedException e) {
            e.printStackTrace();
        }
        return "注册用户: " + userName + " success! ";
    }

}
```

当用户自定义拦截方法时需要实现 MethodInterceptor 接口的 invoke 方法，其使用方式与 Spring 框架中的 AOP 非常相似，也是使用 invocation.proceed 包装，并在 finally 中添加监控信息。

3. 单元测试

整个实例测试了 AOP 与 Spring 框架结合之前的核心代码，包括什么是目标对象、如何组装代理信息、如何调用代理对象。AdvisedSupport 包装了目标对象、用户实现的拦截方法及方法匹配表达式。然后分别通过 JdkDynamicAopProxy、Cglib2AopProxy 两种不同的方法实现代理类，查看是否可以成功拦截方法。

测试结果如下。

```
监控 - Begin By AOP
方法名称：public abstract java.lang.String cn.bugstack.springframework.test.bean.
IUserService.queryUserInfo()
方法耗时：86ms
监控 - End

测试结果：小傅哥，100001，深圳
监控 - Begin By AOP
方法名称：public java.lang.String cn.bugstack.springframework.test.bean.UserService.
register(java.lang.String)
方法耗时：97ms
监控 - End

测试结果：注册用户：花花 success！

Process finished with exit code 0
```

与 AOP 功能的定义一样，通过这样的代理方式、方法匹配和拦截，用户可以在对应的目标方法下执行拦截操作，并输出监控信息。

11.5 本章总结

本章使用 Proxy#newProxyInstance、MethodInterceptor#invoke 验证切面的核心原理，再将功能拆解到 Spring 框架实现中，从中可以看到一个看似复杂但核心内容并不太复杂的技术。因为需要满足后续更多的扩展，所以需要对类进行职责解耦和包装。我们通过使用设计模式，简化调用方法，也可以不断地按需迭代设计模式。

AOP 的功能实现目前还没有与 Spring 结合，只是对切面技术的一个具体实现，这样我们可以先学习如何处理代理对象、过滤方法、拦截方法，体会使用 Cglib 和 JDK 代理的区别。切面技术不仅是在 Spring 框架中有所体现，还会用在其他各类需要减少人工硬编码的场景下，如 RPC、MyBatis、MQ、分布式任务等。

一些核心技术的使用都具有很强的关联性，而且它们都不是孤立存在的。在将整个技术栈串联起来的过程中，我们需要不断地学习、积累，由点到面理解，只有这样才能将一个知识点拓展到知识面和知识体系。

第 12 章
把 AOP 融入 Bean 的生命周期

在 Spring 框架的实现方面，随着内容的扩充和深入，读者会感觉越来越难。但是在经过认真地学习和动手实践，完成一个个代码片段功能的开发和调试后，读者会感觉越来越容易了。

每个读者的知识储备及学习方式，决定了能否把一个较大的项目涉及的功能模块清楚地拆解和组合出来。Spring 框架的实现过程就像搭积木，需要不断地增加新的功能片段并逐步完善。如果在这个过程中缺少了哪些知识，则可以进行补充学习。例如，AOP 实现中涉及的 JDK 代理和 Cglib 代理等，决定了我们能否顺利完成接下来的内容。

- 本章难度：★★★★★
- 本章重点：通过代理工厂、切面拦截调用和切点表达式，借助对象实例化扩展，将代理自动化操作整合到容器中进行管理，以此实现 AOP 的切面功能。

12.1 AOP 与框架整合思考

第 11 章通过基于 Proxy.newProxyInstance 代理操作的方法拦截器和方法匹配器，对匹配的对象进行自定义处理。同时，将核心内容拆解到 Spring 框架中，用于实现 AOP 部分。拆分后基本可以明确各个类的职责，包括代理目标对象属性、拦截器属性、方法匹配属性，以及两种不同的代理操作方式——JDK 和 Cglib。

在实现 AOP 的核心功能之后，我们可以通过单元测试的方式验证切面功能对方法的拦截。一个面向用户使用的功能不太可能让用户操作变复杂，况且没有与 Spring 框架结合的 AOP 也没有太大意义。

因此，本章会整合 AOP 的核心功能与 Spring 框架，最终实现以 Spring 框架配置的方式完成切面的操作。

12.2 AOP 切面设计

在实现 AOP 的核心功能后，将 AOP 切面设计功能融入 Spring 框架中时，需要解决几个问题，如借助 BeanPostProcessor 把动态代理融入 Bean 的生命周期中，组装各项切点、拦截、前置的功能和适配对应的代理器。整体设计结构如图 12-1 所示。

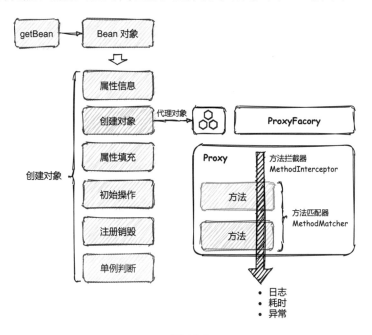

图 12-1

在创建对象的过程中，需要将 XML 文件中配置的代理对象（切面中的一些类对象）实例化，会用到 BeanPostProcessor 类提供的方法来修改 Bean 对象执行初始化前后的扩展信息。但这里需要结合 BeanPostProcessor 实现新的接口和实现类，这样才能定向获取对应的类信息。

与创建之前流程中的普通对象不同，代理对象的创建需要优先于其他对象的创建。在实际开发过程中，需要在 AbstractAutowireCapableBeanFactory#createBean 中优先判断 Bean 对象是否需要代理，如果需要，则直接返回代理对象。在 Spring 的源码中，会对 createBean 和 doCreateBean 方法进行拆分。

设计 AOP 切面需要实现方法拦截器的具体功能，如 BeforeAdvice、AfterAdvice，让用户可以更简单地使用切面功能；还需要整合包装切面表达式及拦截方法，提供不同类型的代理方式的代理工厂，用来包装切面服务。

12.3 AOP 切面实现

1. 工程结构

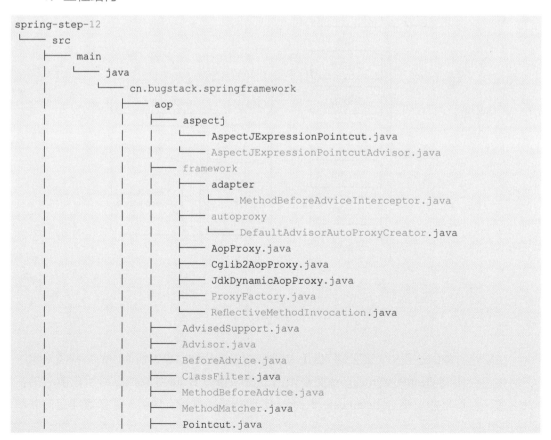

第 12 章 把 AOP 融入 Bean 的生命周期

```
|   |   |       ├── PointcutAdvisor.java
|   |   |       └── TargetSource.java
|   |   ├── beans
|   |   |   ├── factory
|   |   |   |   ├── config
|   |   |   |   |   ├── AutowireCapableBeanFactory.java
|   |   |   |   |   ├── BeanDefinition.java
|   |   |   |   |   ├── BeanFactoryPostProcessor.java
|   |   |   |   |   ├── BeanPostProcessor.java
|   |   |   |   |   ├── BeanReference.java
|   |   |   |   |   ├── ConfigurableBeanFactory.java
|   |   |   |   |   ├── InstantiationAwareBeanPostProcessor.java
|   |   |   |   |   └── SingletonBeanRegistry.java
|   |   |   |   ├── support
|   |   |   |   |   ├── AbstractAutowireCapableBeanFactory.java
|   |   |   |   |   ├── AbstractBeanDefinitionReader.java
|   |   |   |   |   ├── AbstractBeanFactory.java
|   |   |   |   |   ├── BeanDefinitionReader.java
|   |   |   |   |   ├── BeanDefinitionRegistry.java
|   |   |   |   |   ├── CglibSubclassingInstantiationStrategy.java
|   |   |   |   |   ├── DefaultListableBeanFactory.java
|   |   |   |   |   ├── DefaultSingletonBeanRegistry.java
|   |   |   |   |   ├── DisposableBeanAdapter.java
|   |   |   |   |   ├── FactoryBeanRegistrySupport.java
|   |   |   |   |   ├── InstantiationStrategy.java
|   |   |   |   |   └── SimpleInstantiationStrategy.java
|   |   |   |   ├── xml
|   |   |   |   |   └── XmlBeanDefinitionReader.java
|   |   |   |   ├── Aware.java
|   |   |   |   ├── BeanClassLoaderAware.java
|   |   |   |   ├── BeanFactory.java
|   |   |   |   ├── BeanFactoryAware.java
|   |   |   |   ├── BeanNameAware.java
|   |   |   |   ├── ConfigurableListableBeanFactory.java
|   |   |   |   ├── DisposableBean.java
|   |   |   |   ├── FactoryBean.java
|   |   |   |   ├── HierarchicalBeanFactory.java
|   |   |   |   ├── InitializingBean.java
|   |   |   |   └── ListableBeanFactory.java
|   |   |   ├── BeansException.java
|   |   |   ├── PropertyValue.java
|   |   |   └── PropertyValues.java
|   |   ├── context
|   |   |   ├── event
```

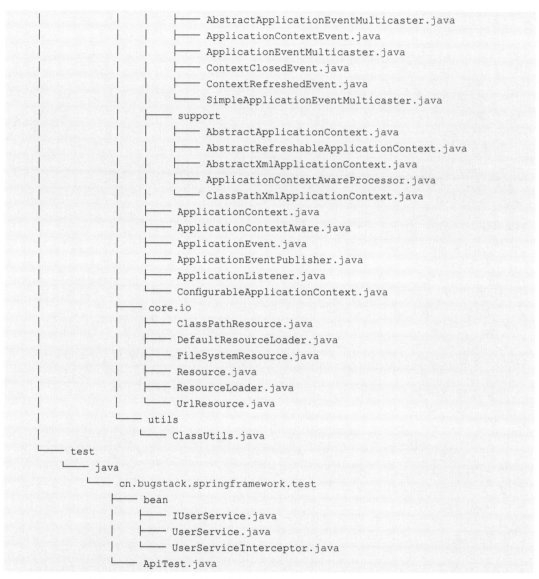

将 AOP 动态代理融入 Bean 的生命周期中类的关系如图 12-2 所示。

从整个类关系图中可以看到，在 BeanPostProcessor 接口实现继承的 InstantiationAwareBeanPostProcessor 接口后，创建了一个自动代理的 DefaultAdvisorAutoProxyCreator 类，这个类负责将整个 AOP 代理融入 Bean 的生命周期中。

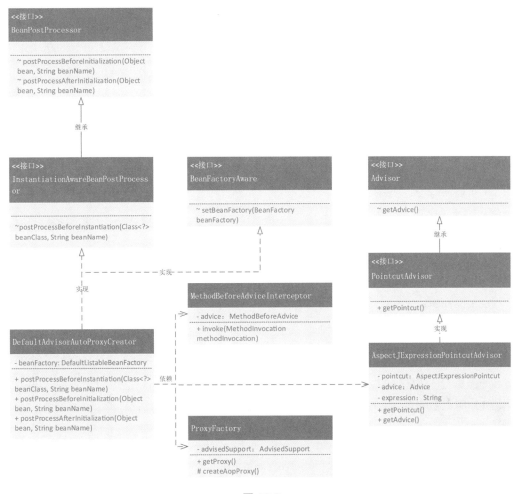

图 12-2

DefaultAdvisorAutoProxyCreator 类依赖于拦截器、代理工厂，以及 Pointcut 与 Advisor 的包装服务 AspectJExpressionPointcutAdvisor，并提供切面、拦截方法和表达式。

Spring 框架中的 AOP 把 Advice 细化为 BeforeAdvice、AfterAdvice、AfterReturningAdvice、ThrowsAdvice，在目前的测试实例中我们只使用了 BeforeAdvice。针对这部分内容，用户可以对比 Spring 的源码进行补充测试。

2. 定义 Advice 拦截器链路

源码详见：cn.bugstack.springframework.aop.BeforeAdvice。

```
public interface BeforeAdvice extends Advice {

}
```

源码详见：cn.bugstack.springframework.aop.MethodBeforeAdvice。

```
public interface MethodBeforeAdvice extends BeforeAdvice {

    /**
     * Callback before a given method is invoked.
     *
     * @param method method being invoked
     * @param args   arguments to the method
     * @param target target of the method invocation. May be <code>null</code>.
     * @throws Throwable if this object wishes to abort the call.
     *                   Any exception thrown will be returned to the caller if it's
     *                   allowed by the method signature. Otherwise the exception
     *                   will be wrapped as a runtime exception.
     */
    void before(Method method, Object[] args, Object target) throws Throwable;

}
```

在 Spring 框架中，Advice 是通过方法拦截器 MethodInterceptor 实现的，即围绕 Advice 做一个类似拦截器的链路——Before Advice、After advice 等。因为暂时不需要太多接口，所以只定义了一个 MethodBeforeAdvice 接口。

3. 定义 Advisor 访问者

源码详见：cn.bugstack.springframework.aop.Advisor。

```
public interface Advisor {

    /**
     * Return the advice part of this aspect. An advice may be an
     * interceptor, a before advice, a throws advice, etc.
     * @return the advice that should apply if the pointcut matches
     * @see org.aopalliance.intercept.MethodInterceptor
     * @see BeforeAdvice
     */
    Advice getAdvice();

}
```

源码详见：cn.bugstack.springframework.aop.PointcutAdvisor。

```java
public interface PointcutAdvisor extends Advisor {

    /**
     * Get the Pointcut that drives this advisor.
     */
    Pointcut getPointcut();

}
```

Advisor 是 Pointcut 功能和 Advice 功能的组合。Pointcut 用于获取 JoinPoint，而 Advice 取决于 JoinPoint 执行什么操作。

源码详见：cn.bugstack.springframework.aop.aspectj.AspectJExpressionPointcutAdvisor。

```java
public class AspectJExpressionPointcutAdvisor implements PointcutAdvisor {

    // 切面
    private AspectJExpressionPointcut pointcut;
    // 具体的拦截方法
    private Advice advice;
    // 表达式
    private String expression;

    public void setExpression(String expression){
        this.expression = expression;
    }

    @Override
    public Pointcut getPointcut() {
        if (null == pointcut) {
            pointcut = new AspectJExpressionPointcut(expression);
        }
        return pointcut;
    }

    @Override
    public Advice getAdvice() {
        return advice;
    }

    public void setAdvice(Advice advice){
        this.advice = advice;
    }

}
```

AspectJExpressionPointcutAdvisor 实现了 PointcutAdvisor 接口，将切面 pointcut、拦截方法 advice 和具体的拦截表达式包装在一起。这样就可以在 XML 文件的配置中定义一个 PointcutAdvisor 切面拦截器。

4. 方法拦截器

源码详见：cn.bugstack.springframework.aop.framework.adapter.MethodBeforeAdviceInterceptor。

```
public class MethodBeforeAdviceInterceptor implements MethodInterceptor {

    private MethodBeforeAdvice advice;

    public MethodBeforeAdviceInterceptor(MethodBeforeAdvice advice) {
        this.advice = advice;
    }

    @Override
    public Object invoke(MethodInvocation methodInvocation) throws Throwable {
        this.advice.before(methodInvocation.getMethod(), methodInvocation.getArguments(), methodInvocation.getThis());
        return methodInvocation.proceed();
    }

}
```

通过 MethodBeforeAdviceInterceptor 实现了 MethodInterceptor 接口，在 invoke 方法中调用 advice 中的 before 方法，传入对应的参数信息。

通过 advice.before 方法实现 MethodBeforeAdvice 接口，并进行了相应的处理。可以看到，MethodInterceptor 实现类的功能与之前的测试是一样的，只不过现在交给了 Spring 框架来处理。

5. 代理工厂

源码详见：cn.bugstack.springframework.aop.framework.ProxyFactory。

```
public class ProxyFactory {

    private AdvisedSupport advisedSupport;

    public ProxyFactory(AdvisedSupport advisedSupport) {
        this.advisedSupport = advisedSupport;
    }
```

```java
public Object getProxy() {
    return createAopProxy().getProxy();
}

private AopProxy createAopProxy() {
    if (advisedSupport.isProxyTargetClass()) {
        return new Cglib2AopProxy(advisedSupport);
    }

    return new JdkDynamicAopProxy(advisedSupport);
}
```

这个代理工厂主要解决了选择 JDK 和 Cglib 两种代理的问题。有了代理工厂，用户就可以按照不同的创建需求进行控制。

6. 融入 Bean 生命周期的自动代理创建者

源码详见：cn.bugstack.springframework.aop.framework.autoproxy.DefaultAdvisorAutoProxyCreator。

```java
public class DefaultAdvisorAutoProxyCreator implements InstantiationAwareBeanPostProcessor, BeanFactoryAware {

    private DefaultListableBeanFactory beanFactory;

    @Override
    public void setBeanFactory(BeanFactory beanFactory) throws BeansException {
        this.beanFactory = (DefaultListableBeanFactory) beanFactory;
    }

    @Override
    public Object postProcessBeforeInstantiation(Class<?> beanClass, String beanName) throws BeansException {

        if (isInfrastructureClass(beanClass)) return null;

        Collection<AspectJExpressionPointcutAdvisor> advisors = beanFactory.getBeansOfType(AspectJExpressionPointcutAdvisor.class).values();

        for (AspectJExpressionPointcutAdvisor advisor : advisors) {
            ClassFilter classFilter = advisor.getPointcut().getClassFilter();
            if (!classFilter.matches(beanClass)) continue;
```

```java
            AdvisedSupport advisedSupport = new AdvisedSupport();

            TargetSource targetSource = null;
            try {
                targetSource = new TargetSource(beanClass.getDeclaredConstructor().newInstance());
            } catch (Exception e) {
                e.printStackTrace();
            }
            advisedSupport.setTargetSource(targetSource);
            advisedSupport.setMethodInterceptor((MethodInterceptor) advisor.getAdvice());
            advisedSupport.setMethodMatcher(advisor.getPointcut().getMethodMatcher());
            advisedSupport.setProxyTargetClass(false);

            return new ProxyFactory(advisedSupport).getProxy();

        }

        return null;
    }
}
```

DefaultAdvisorAutoProxyCreator 类的主要核心实现在 postProcessBeforeInstantiation 方法中，从 beanFactory.getBeansOfType 获取 AspectJExpressionPointcutAdvisor 开始。

在获取了 advisors 后，首先遍历相应的 AspectJExpressionPointcutAdvisor 来填充对应的属性信息（如目标对象、拦截方法、匹配器），然后返回代理对象。

现在，调用方获取的 Bean 对象就是一个已经被切面注入的对象，当调用方法时，其会被按需拦截，处理用户需要的信息。

12.4 切面使用测试

1. 事先准备

```java
public class UserService implements IUserService {

    public String queryUserInfo() {
        try {
```

```
            Thread.sleep(new Random(1).nextInt(100));
        } catch (InterruptedException e) {
            e.printStackTrace();
        }
        return "小傅哥, 100001, 深圳 ";
    }

    public String register(String userName) {
        try {
            Thread.sleep(new Random(1).nextInt(100));
        } catch (InterruptedException e) {
            e.printStackTrace();
        }
        return "注册用户: " + userName + " success！ ";
    }

}
```

UserService 提供了两个不同的方法，还可以增加新的类并将其加入测试。在后面的测试过程中，会对这两个方法添加切面拦截处理，以及输出方法执行耗时。

2. 自定义拦截方法

```
public class UserServiceBeforeAdvice implements MethodBeforeAdvice {

    @Override
    public void before(Method method, Object[] args, Object target) throws Throwable {
        System.out.println("拦截方法: " + method.getName());
    }

}
```

与第 11 章的拦截方法相比，这里不再是实现 MethodInterceptor 接口，而是实现 MethodBeforeAdvice 环绕拦截。在 MethodBeforeAdvice 方法中，可以获取方法的一些信息。如果还开发了 MethodAfterAdvice 接口，则可以同时实现 MethodInterceptor 和 MethodAfterAdvice 两个接口。

3. 在 spring.xml 配置文件中配置 AOP

```
<beans>

    <bean id="userService" class="cn.bugstack.springframework.test.bean.UserService"/>

    <bean class="cn.bugstack.springframework.aop.framework.autoproxy.
```

```xml
DefaultAdvisorAutoProxyCreator"/>

    <bean id="beforeAdvice" class="cn.bugstack.springframework.test.bean.
UserServiceBeforeAdvice"/>

    <bean id="methodInterceptor" class="cn.bugstack.springframework.aop.framework.
adapter.MethodBeforeAdviceInterceptor">
        <property name="advice" ref="beforeAdvice"/>
    </bean>

    <bean id="pointcutAdvisor" class="cn.bugstack.springframework.aop.aspectj.
AspectJExpressionPointcutAdvisor">
        <property name="expression" value="execution(* cn.bugstack.springframework.
test.bean.IUserService.*(..))"/>
        <property name="advice" ref="methodInterceptor"/>
    </bean>

</beans>
```

用户可以在 spring.xml 配置文件中配置 AOP，因为已经将 AOP 的功能融入 Bean 的生命周期中，所以新增的拦截方法都会被自动处理。

4. 单元测试

```
@Test
public void test_aop() {
    ClassPathXmlApplicationContext applicationContext = new ClassPathXmlApplicationContext
("classpath:spring.xml");
    IUserService userService = applicationContext.getBean("userService", IUserService.class);
    System.out.println("测试结果: " + userService.queryUserInfo());
}
```

在单元测试中，用户只需要按照正常的方式获取和使用 Bean 对象。如果被切面拦截了，则获取的就是对应的代理对象的处理结果。

测试结果如图 12-3 所示。

```
拦截方法：queryUserInfo
测试结果：小傅哥, 100001, 深圳

Process finished with exit code 0
```

从测试结果中可以看到，拦截方法已经生效，不需要用户手动处理切面、拦截方法等。从截图中可以看到，这时的 IUserService 就是一个代理对象。

图 12-3

12.5　本章总结

　　本章实现的 AOP 功能主要是将在单元测试中的切面拦截交给 Spring 框架中的 XML 文件来配置，不需要用户手动处理。对于相应的功能如何与 Spring 框架的 Bean 生命周期结合起来，本章使用的是 BeanPostProcessor，因为它可以用于修改 Bean 对象执行初始化方法之前新实例化 Bean 对象的扩展点。

　　一个功能的实现往往包括核心部分、组装部分和链接部分。为了明确各自的职责，需要创建接口和类，由不同关系的继承、实现进行组装。开发人员只有明确了各自的职责，才能灵活地扩展相应的功能逻辑，否则很难进行大型系统的开发和创建。

　　目前，我们实现的 AOP 与 Spring 源码中的核心逻辑是类似的，但 AOP 更简单一些，而且没有考虑在更多的复杂场景中遇到的问题，如是否有构造函数、是否为代理中的切面等。同时，只要是 Java 中的一些特性，都需要在 Spring 源码中进行完整的实现，否则使用时就会遇到各种问题。

第 13 章
自动扫描注册 Bean 对象

在开发项目时，通常会渐进式地完成需求，使产品功能从能用到好用。最开始只是开发出来一些命令调用、API 步骤接口等功能，多数需要用户进行逻辑处理。如果想要使产品从能用过渡到好用，则需要使用一些设计模式来屏蔽功能细节，让用户只关心需要的内容。

在 Spring 框架的实现过程中，从产生 Spring Bean 容器到配置注册 Bean 对象，可以满足基本的使用需求。但在实际中，需要配置大量的 Bean 对象。从用户的角度来看，这是一件重复且烦琐的事情，因此需要对注册流程进行优化和完善，提升用户的产品体验。

- 本章难度：★★★☆☆
- 本章重点：定义属性标识、对象的注解方式，当容器扫描 XML 配置时，提取 Bean 对象的属性和对象信息，将其注册到 Bean 对象的定义组中。通过这部分的自动化扫描衔接，优化对象的注册流程。

13.1 注入对象完善点

本章将介绍 IOC 和 AOP 的核心内容。在 Spring 框架的早期版本中，这些功能需要在 spring.xml 配置文件中配置。这与目前实际使用的 Spring 框架还有很大差别，之后对核心功能逻辑进行完善，利用更少的配置实现更简化的操作，包括包路径的扫描、注解配置的使用、占位符属性的填充等。我们的目标就是在目前的核心逻辑上填充一些自动化功能。在此过程中，我们可以学习 IOC 和 AOP 的设计与实现，体会代码逻辑的实现过程，积累编程经验。

13.2 自动扫描注册设计

为了简化 Bean 对象的配置，使整个 Bean 对象的注册通过自动扫描完成，需要完善功能，包括：扫描路径入口、XML 解析扫描信息、给需要扫描的 Bean 对象做注解标记、扫描 Class 类提取 Bean 对象注册的基本信息、组装注册信息并注册成 Bean 对象。在可以实现自定义注解和配置扫描路径的情况下，完成 Bean 对象的注册。此外，还需要解决一个配置中占位符属性的问题，如通过占位符属性值 ${token} 给 Bean 对象注入属性信息。这个操作需要使用 BeanFactoryPostProcessor 类，它可以在所有的 BeanDefinition 加载完成后，将 Bean 对象实例化之前，提供修改 BeanDefinition 属性的机制。实现这部分内容是为了后续将此类内容扩展到自动化配置处理中。整体的设计结构如图 13-1 所示。

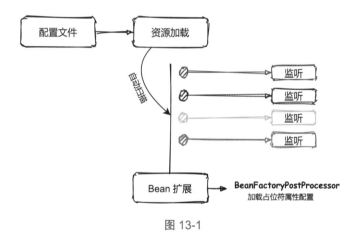

图 13-1

结合 Bean 对象的生命周期来看，包扫描只不过是扫描特定注解的类，提取类的相关信息并组装成 BeanDefinition，注册到容器中。

在 XmlBeanDefinitionReader 类中解析标签，首先扫描类组件 BeanDefinition，然后通过 ClassPathBeanDefinitionScanner#doScan 方法注册到 Spring Bean 容器中。

自动扫描注册主要是扫描添加了自定义注解的类，在 XML 加载过程中提取类的信息，组装 BeanDefinition 并注册到 Spring Bean 容器中。因此，会使用配置包路径，在 XmlBeanDefinitionReader 类中解析并做相应的处理，包括对类的扫描、获取注解信息等，还包括 BeanFactoryPostProcessor 的使用。因为需要完成对占位符配置信息的加载，所以在加载完成所有的 BeanDefinition 后，实例化 Bean 对象之前，使用 BeanFactoryPostProcessor 来修改 BeanDefinition 的属性信息。需要注意的是，这部分的实现也是为后续将占位符配置到注解上做准备。

13.3　自动扫描注册实现

1. 工程结构

```
spring-step-13
└── src
    ├── main
    │   └── java
    │       └── cn.bugstack.springframework
    │           ├── aop
    │           │   ├── aspectj
    │           │   │   ├── AspectJExpressionPointcut.java
    │           │   │   └── AspectJExpressionPointcutAdvisor.java
    │           │   ├── framework
    │           │   │   ├── adapter
    │           │   │   │   └── MethodBeforeAdviceInterceptor.java
    │           │   │   ├── autoproxy
    │           │   │   │   └── DefaultAdvisorAutoProxyCreator.java
    │           │   │   ├── AopProxy.java
    │           │   │   ├── Cglib2AopProxy.java
    │           │   │   ├── JdkDynamicAopProxy.java
    │           │   │   ├── ProxyFactory.java
    │           │   │   └── ReflectiveMethodInvocation.java
    │           │   ├── AdvisedSupport.java
    │           │   ├── Advisor.java
    │           │   ├── BeforeAdvice.java
    │           │   ├── ClassFilter.java
    │           │   ├── MethodBeforeAdvice.java
    │           │   ├── MethodMatcher.java
    │           │   ├── Pointcut.java
    │           │   ├── PointcutAdvisor.java
    │           │   └── TargetSource.java
```

第 13 章 自动扫描注册 Bean 对象

```
|   |   ├── beans
|   |   │   ├── factory
|   |   │   │   ├── config
|   |   │   │   │   ├── AutowireCapableBeanFactory.java
|   |   │   │   │   ├── BeanDefinition.java
|   |   │   │   │   ├── BeanFactoryPostProcessor.java
|   |   │   │   │   ├── BeanPostProcessor.java
|   |   │   │   │   ├── BeanReference.java
|   |   │   │   │   ├── ConfigurableBeanFactory.java
|   |   │   │   │   ├── InstantiationAwareBeanPostProcessor.java
|   |   │   │   │   └── SingletonBeanRegistry.java
|   |   │   │   ├── support
|   |   │   │   │   ├── AbstractAutowireCapableBeanFactory.java
|   |   │   │   │   ├── AbstractBeanDefinitionReader.java
|   |   │   │   │   ├── AbstractBeanFactory.java
|   |   │   │   │   ├── BeanDefinitionReader.java
|   |   │   │   │   ├── BeanDefinitionRegistry.java
|   |   │   │   │   ├── CglibSubclassingInstantiationStrategy.java
|   |   │   │   │   ├── DefaultListableBeanFactory.java
|   |   │   │   │   ├── DefaultSingletonBeanRegistry.java
|   |   │   │   │   ├── DisposableBeanAdapter.java
|   |   │   │   │   ├── FactoryBeanRegistrySupport.java
|   |   │   │   │   ├── InstantiationStrategy.java
|   |   │   │   │   └── SimpleInstantiationStrategy.java
|   |   │   │   ├── xml
|   |   │   │   │   └── XmlBeanDefinitionReader.java
|   |   │   │   ├── Aware.java
|   |   │   │   ├── BeanClassLoaderAware.java
|   |   │   │   ├── BeanFactory.java
|   |   │   │   ├── BeanFactoryAware.java
|   |   │   │   ├── BeanNameAware.java
|   |   │   │   ├── ConfigurableListableBeanFactory.java
|   |   │   │   ├── DisposableBean.java
|   |   │   │   ├── FactoryBean.java
|   |   │   │   ├── HierarchicalBeanFactory.java
|   |   │   │   ├── InitializingBean.java
|   |   │   │   ├── ListableBeanFactory.java
|   |   │   │   └── PropertyPlaceholderConfigurer.java
|   |   │   ├── BeansException.java
|   |   │   ├── PropertyValue.java
|   |   │   └── PropertyValues.java
|   |   ├── context
|   |   │   ├── annotation
|   |   │   │   ├── ClassPathBeanDefinitionScanner.java
```

```
|   |   |           ├── ClassPathScanningCandidateComponentProvider.java
|   |   |           └── Scope.java
|   |   ├── event
|   |   |   ├── AbstractApplicationEventMulticaster.java
|   |   |   ├── ApplicationContextEvent.java
|   |   |   ├── ApplicationEventMulticaster.java
|   |   |   ├── ContextClosedEvent.java
|   |   |   ├── ContextRefreshedEvent.java
|   |   |   └── SimpleApplicationEventMulticaster.java
|   |   ├── support
|   |   |   ├── AbstractApplicationContext.java
|   |   |   ├── AbstractRefreshableApplicationContext.java
|   |   |   ├── AbstractXmlApplicationContext.java
|   |   |   ├── ApplicationContextAwareProcessor.java
|   |   |   └── ClassPathXmlApplicationContext.java
|   |   ├── ApplicationContext.java
|   |   ├── ApplicationContextAware.java
|   |   ├── ApplicationEvent.java
|   |   ├── ApplicationEventPublisher.java
|   |   ├── ApplicationListener.java
|   |   └── ConfigurableApplicationContext.java
|   ├── core.io
|   |   ├── ClassPathResource.java
|   |   ├── DefaultResourceLoader.java
|   |   ├── FileSystemResource.java
|   |   ├── Resource.java
|   |   ├── ResourceLoader.java
|   |   └── UrlResource.java
|   ├── stereotype
|   |   └── Component.java
|   └── utils
|       └── ClassUtils.java
└── test
    └── java
        └── cn.bugstack.springframework.test
            ├── bean
            |   ├── IUserService.java
            |   └── UserService.java
            └── ApiTest.java
```

在 Bean 的生命周期中自动加载包扫描、注册 Bean 对象和配置占位符属性的类的关系如图 13-2 所示。

第 13 章 自动扫描注册 Bean 对象

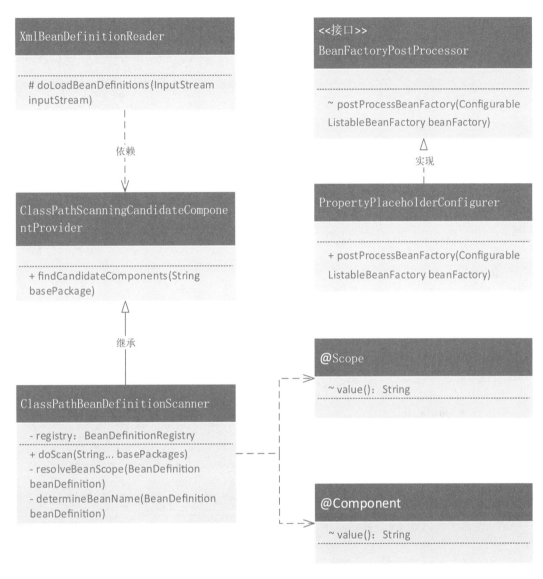

图 13-2

整个类涉及的内容并不多，主要包括 XML 解析类 XmlBeanDefinitionReader 对 ClassPathBeanDefinitionScanner#doScan 方法的使用。首先，在 doScan 方法中处理所有指定路径下添加的注解类，拆解出类的名称、作用范围等；然后，创建 BeanDefinition，以便用于注册 Bean 对象。目前，PropertyPlaceholderConfigurer 看上去像单独的内容，后续会把它与自动加载 Bean 对象进行整合，在注解上使用占位符配置一些配置文件中的属性信息。

2. 处理占位符配置

源码详见：cn.bugstack.springframework.beans.factory.PropertyPlaceholderConfigurer。

```java
public class PropertyPlaceholderConfigurer implements BeanFactoryPostProcessor {

    /**
     * Default placeholder prefix: {@value}
     */
    public static final String DEFAULT_PLACEHOLDER_PREFIX = "${";

    /**
     * Default placeholder suffix: {@value}
     */
    public static final String DEFAULT_PLACEHOLDER_SUFFIX = "}";

    private String location;

    @Override
    public void postProcessBeanFactory(ConfigurableListableBeanFactory beanFactory) throws BeansException {
        // 加载属性文件
        try {
            DefaultResourceLoader resourceLoader = new DefaultResourceLoader();
            Resource resource = resourceLoader.getResource(location);
            Properties properties = new Properties();
            properties.load(resource.getInputStream());

            String[] beanDefinitionNames = beanFactory.getBeanDefinitionNames();
            for (String beanName : beanDefinitionNames) {
                BeanDefinition beanDefinition = beanFactory.getBeanDefinition(beanName);

                PropertyValues propertyValues = beanDefinition.getPropertyValues();
                for (PropertyValue propertyValue : propertyValues.getPropertyValues()) {
                    Object value = propertyValue.getValue();
                    if (!(value instanceof String)) continue;
                    String strVal = (String) value;
                    StringBuilder buffer = new StringBuilder(strVal);
                    int startIdx = strVal.indexOf(DEFAULT_PLACEHOLDER_PREFIX);
                    int stopIdx = strVal.indexOf(DEFAULT_PLACEHOLDER_SUFFIX);
                    if (startIdx != -1 && stopIdx != -1 && startIdx < stopIdx) {
                        String propKey = strVal.substring(startIdx + 2, stopIdx);
                        String propVal = properties.getProperty(propKey);
                        buffer.replace(startIdx, stopIdx + 1, propVal);
```

```
                        propertyValues.addPropertyValue(new PropertyValue(propertyValue.
getName(), buffer.toString()));
                    }
                }
            }
        } catch (IOException e) {
            throw new BeansException("Could not load properties", e);
        }
    }

    public void setLocation(String location) {
        this.location = location;
    }

}
```

根据 BeanFactoryPostProcessor 接口在 Bean 生命周期中的属性特点，可以在 Bean 对象实例化之前改变属性信息。这里通过实现 BeanFactoryPostProcessor 接口，完成对配置文件的加载，以及获取占位符在属性文件中的配置。这样就可以把获取的配置信息放置到属性配置中，即 buffer.replace(startIdx, stopIdx + 1, propVal); propertyValues.addPropertyValue。

3. 定义拦截注解

源码详见：cn.bugstack.springframework.context.annotation.Scope。

```
@Target({ElementType.TYPE, ElementType.METHOD})
@Retention(RetentionPolicy.RUNTIME)
@Documented
public @interface Scope {

    String value() default "singleton";

}
```

自定义注解用于配置作用域，在配置 Bean 对象注解时，方便获取 Bean 对象的作用域。需要注意的是，一般都使用默认的 singleton。

源码详见：cn.bugstack.springframework.stereotype.Component。

```
@Target(ElementType.TYPE)
@Retention(RetentionPolicy.RUNTIME)
@Documented
public @interface Component {
```

```
    String value() default "";
}
```

对于配置 Class 类，除了使用 Component 进行自定义注解，还可以使用 Service、Controller，它们的处理方式基本一致，这里只介绍使用 Component 进行自定义注释的方法。

4. 处理对象扫描装配

源码详见：cn.bugstack.springframework.context.annotation.ClassPathScanningCandidateComponentProvider。

```
public class ClassPathScanningCandidateComponentProvider {

    public Set<BeanDefinition> findCandidateComponents(String basePackage) {
        Set<BeanDefinition> candidates = new LinkedHashSet<>();
        Set<Class<?>> classes = ClassUtil.scanPackageByAnnotation(basePackage, Component.class);
        for (Class<?> clazz : classes) {
            candidates.add(new BeanDefinition(clazz));
        }
        return candidates;
    }

}
```

首先提供一个可以通过配置路径 basePackage=cn.bugstack.springframework.test.bean 解析出 classes 信息的方法 findCandidateComponents，然后通过这个方法就可以扫描所有使用 Component 注解的 Bean 对象。

源码详见：cn.bugstack.springframework.context.annotation.ClassPathBeanDefinitionScanner。

```
public class ClassPathBeanDefinitionScanner extends ClassPathScanningCandidateComponentProvider {

    private BeanDefinitionRegistry registry;

    public ClassPathBeanDefinitionScanner(BeanDefinitionRegistry registry) {
        this.registry = registry;
    }

    public void doScan(String… basePackages) {
        for (String basePackage : basePackages) {
            Set<BeanDefinition> candidates = findCandidateComponents(basePackage);
```

```java
            for (BeanDefinition beanDefinition : candidates) {
                // 解析 Bean 对象的作用域 singleton、prototype
                String beanScope = resolveBeanScope(beanDefinition);
                if (StrUtil.isNotEmpty(beanScope)) {
                    beanDefinition.setScope(beanScope);
                }
                registry.registerBeanDefinition(determineBeanName(beanDefinition), beanDefinition);
            }
        }
    }

    private String resolveBeanScope(BeanDefinition beanDefinition) {
        Class<?> beanClass = beanDefinition.getBeanClass();
        Scope scope = beanClass.getAnnotation(Scope.class);
        if (null != scope) return scope.value();
        return StrUtil.EMPTY;
    }

    private String determineBeanName(BeanDefinition beanDefinition) {
        Class<?> beanClass = beanDefinition.getBeanClass();
        Component component = beanClass.getAnnotation(Component.class);
        String value = component.value();
        if (StrUtil.isEmpty(value)) {
            value = StrUtil.lowerFirst(beanClass.getSimpleName());
        }
        return value;
    }
}
```

ClassPathBeanDefinitionScanner 是继承自 ClassPathScanningCandidateComponentProvider 的具体扫描包处理的类。在 doScan 中，除了需要获取扫描的类信息，还需要获取 Bean 的作用域和类名。

5. 解析 XML 中的调用扫描

源码详见：cn.bugstack.springframework.beans.factory.xml.XmlBeanDefinitionReader。

```java
public class XmlBeanDefinitionReader extends AbstractBeanDefinitionReader {

    protected void doLoadBeanDefinitions(InputStream inputStream) throws ClassNotFoundException, DocumentException {
        SAXReader reader = new SAXReader();
```

```java
        Document document = reader.read(inputStream);
        Element root = document.getRootElement();
        
        // 解析 context:component-scan 标签，扫描包中的类并提取相关信息，用于组装 BeanDefinition
        Element componentScan = root.element("component-scan");
        if (null != componentScan) {
            String scanPath = componentScan.attributeValue("base-package");
            if (StrUtil.isEmpty(scanPath)) {
                throw new BeansException("The value of base-package attribute can not be empty or null");
            }
            scanPackage(scanPath);
        }
        
        // 省略其他代码
        
        // 注册 BeanDefinition
        getRegistry().registerBeanDefinition(beanName, beanDefinition);
    }
    
    private void scanPackage(String scanPath) {
        String[] basePackages = StrUtil.splitToArray(scanPath, ',');
        ClassPathBeanDefinitionScanner scanner = new ClassPathBeanDefinitionScanner(getRegistry());
        scanner.doScan(basePackages);
    }
}
```

XmlBeanDefinitionReader 主要是在加载配置文件后处理新增的自定义配置属性 component-scan 的，解析后调用 scanPackage 方法，其实也就是 ClassPathBeanDefinition Scanner#doScan 功能。需要注意的是，为了方便加载和解析 XML 文件，已经将 XmlBean DefinitionReader 全部替换为 dom4j 的方式进行解析。

13.4 注册 Bean 对象测试

1. 事先准备

```java
@Component("userService")
public class UserService implements IUserService {
```

```java
    private String token;

    public String queryUserInfo() {
        try {
            Thread.sleep(new Random(1).nextInt(100));
        } catch (InterruptedException e) {
            e.printStackTrace();
        }
        return "小傅哥,100001,深圳";
    }

    public String register(String userName) {
        try {
            Thread.sleep(new Random(1).nextInt(100));
        } catch (InterruptedException e) {
            e.printStackTrace();
        }
        return "注册用户: " + userName + " success! ";
    }

    @Override
    public String toString() {
        return "UserService#token = { " + token + " }";
    }

    public String getToken() {
        return token;
    }

    public void setToken(String token) {
        this.token = token;
    }
}
```

给 UserService 类添加一个自定义注解 @Component("userService") 和一个属性信息 token,分别用于测试包扫描和占位符属性。

2. 属性配置文件

```
token=RejDlI78hu223Opo983Ds
```

这里配置一个 token 的属性信息,用于通过占位符的方式获取信息。

3. 在 spring.xml 配置文件中配置对象

（1）spring-property.xml。

```xml
<?xml version="1.0" encoding="UTF-8"?>
<beans xmlns="http://www.springframework.org/schema/beans"
    xmlns:xsi="http://www.w3.org/2001/XMLSchema-instance"
    xmlns:context="http://www.springframework.org/schema/context"
    xsi:schemaLocation="http://www.springframework.org/schema/beans
        http://www.springframework.org/schema/beans/spring-beans.xsd
       http://www.springframework.org/schema/context">

    <bean class="cn.bugstack.springframework.beans.factory.PropertyPlaceholderConfigurer">
        <property name="location" value="classpath:token.properties"/>
    </bean>

    <bean id="userService" class="cn.bugstack.springframework.test.bean.UserService">
        <property name="token" value="${token}"/>
    </bean>

</beans>
```

加载 classpath:token.properties 后，设置占位符属性值 ${token}。

（2）spring-scan.xml。

```xml
<?xml version="1.0" encoding="UTF-8"?>
<beans xmlns="http://www.springframework.org/schema/beans"
    xmlns:xsi="http://www.w3.org/2001/XMLSchema-instance"
    xmlns:context="http://www.springframework.org/schema/context"
    xsi:schemaLocation="http://www.springframework.org/schema/beans
        http://www.springframework.org/schema/beans/spring-beans.xsd
       http://www.springframework.org/schema/context">

    <context:component-scan base-package="cn.bugstack.springframework.test.bean"/>

</beans>
```

添加 component-scan 属性后，设置包扫描根路径。

4. 单元测试（占位符）

```
@Test
public void test_property() {
    ClassPathXmlApplicationContext applicationContext = new
ClassPathXmlApplicationContext("classpath:spring-property.xml");
```

```
    IUserService userService = applicationContext.getBean("userService", IUserService.
class);
    System.out.println("测试结果: " + userService);
}
```

测试结果如下。

```
测试结果: UserService#token = { RejDlI78hu223Opo983Ds }

Process finished with exit code 0
```

从测试结果中可以看到，UserService 中的 token 属性已经通过占位符的方式设置配置文件中的 token.properties 的属性值。

5. 单元测试（包扫描）

```
@Test
public void test_scan() {
    ClassPathXmlApplicationContext applicationContext = new ClassPathXmlApplicationContext
("classpath:spring-scan.xml");
    IUserService userService = applicationContext.getBean("userService", IUserService.
class);
    System.out.println("测试结果: " + userService.queryUserInfo());
}
```

测试结果如下。

```
测试结果: 小傅哥, 100001, 深圳

Process finished with exit code 0
```

从测试结果中可以看到，现在使用注解的方式就可以使 Class 类完成 Bean 对象的注册。

13.5 本章总结

目前，在 Spring 框架中添加的功能都是在 IOC 和 AOP 核心的基础上进行扩展的。这些扩展的功能也是在完善 Bean 的生命周期。

当不断丰富 Spring 的各项功能时，我们也可以将经常使用的 Spring 的一些功能想法融入进来，如 Spring 是如何动态切换数据源的、线程池是如何提供配置的等。

第 14 章
通过注解注入属性信息

在开发本章的功能需求之前，不知道大家是否了解扰动函数、方差稳定性、斐波那契散列法、梅森旋转算法等。虽然很多程序员不了解这些编程之外的数学内容，但是不会影响程序的开发。如果程序的需求是面向大规模的用户，就需要使用很多细节功能完善代码逻辑来支撑用户的使用，而这些细节更多的是来自数学方面的知识，最终的代码就是对数学逻辑的具体落地实现。

本章将继续结合自动扫描注入的内容，将一个 Bean 对象的属性信息通过注解注入。注入的属性信息包括属性值和对象，而这些扩展功能都来自 BeanPostProcessor、Aware 的运用。

- 本章难度：★★★☆☆
- 本章重点：定义属性和对象的标记性注解。在创建对象实例化后，BeanPostProcessor 的实现类通过对象获取和反射的方式对 Bean 对象中含有注解的属性字段进行属性和对象的注入。

14.1 引入注入注解

在 IOC、AOP 两大核心功能模块的支撑下，虽然可以管理 Bean 对象的注册和获取，但是有些难以操作。因此，在第 13 章中，我们将需要在 spring.xml 配置文件中进行手动配置 Bean 对象的操作，更改为可以自动扫描带有注解 @Component 的对象，将 Bean 对象自动装配和注册到 Spring Bean 容器中。

在自动扫描包注册 Bean 对象之后，需要将原来在 spring.xml 配置文件中通过 property name="token" 配置属性和 Bean 的操作更改为自动注入。这就像使用 Spring 框架中的 @Autowired、@Value 注解一样，完成对属性和对象的注解配置的注入操作。

14.2　注入属性信息设计

在完成 Bean 对象的基础功能之后，后面陆续添加的功能都是围绕 Bean 的生命周期展开的，如修改 Bean 的定义 BeanFactoryPostProcessor、处理 Bean 的属性需要使用 BeanPostProcessor、完成一些额外定义的属性操作需要专门继承 BeanPostProcessor 提供的接口，只有这样才能通过 instanceof 判断出具有标记性的接口。因此，关于 Bean 的操作，以及监听 Aware、获取 BeanFactory，都需要在 Bean 的生命周期中完成。在设计属性和注入 Bean 对象时，使用 BeanPostProcessor 修改属性值。整体设计结构如图 14-1 所示。

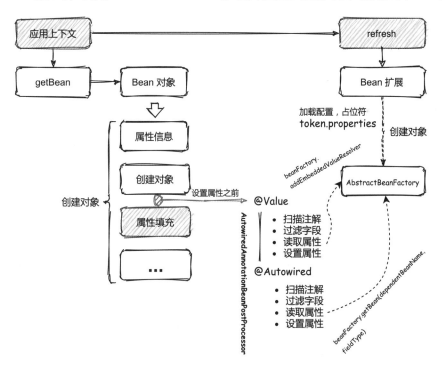

图 14-1

如果想要处理自动扫描注入（包括属性注入、对象注入），就需要在填充对象属性 applyPropertyValues 之前，把属性信息写入 PropertyValues 集合。这一步操作相当于解决了以前在 spring.xml 配置文件中注入配置属性的过程。

在读取属性时，需要通过扫描 Bean 对象的属性字段，判断是否配置了 @Value 注解。并把扫描到的配置了 @Value 注解的属性填充对应的属性信息。需要注意的是，属性的配置信息需要依赖 BeanFactoryPostProcessor 的实现类 PropertyPlaceholderConfigurer，把值写入 AbstractBeanFactory 的 embeddedValueResolvers 集合中，这样才能在填充属性时利用 beanFactory 获取相应的属性值。

@Autowired 注解对对象的注入，其实与属性注入的唯一区别是对象的获取：beanFactory.getBean(fieldType)。

在所有属性被注入 PropertyValues 集合之后进行属性填充，此时就会把获取的配置和对象填充到属性上，也就实现了自动注入的功能。

14.3 注入属性信息实现

1. 工程结构

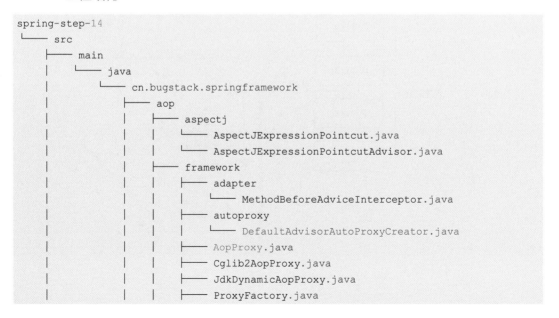

第 14 章 通过注解注入属性信息

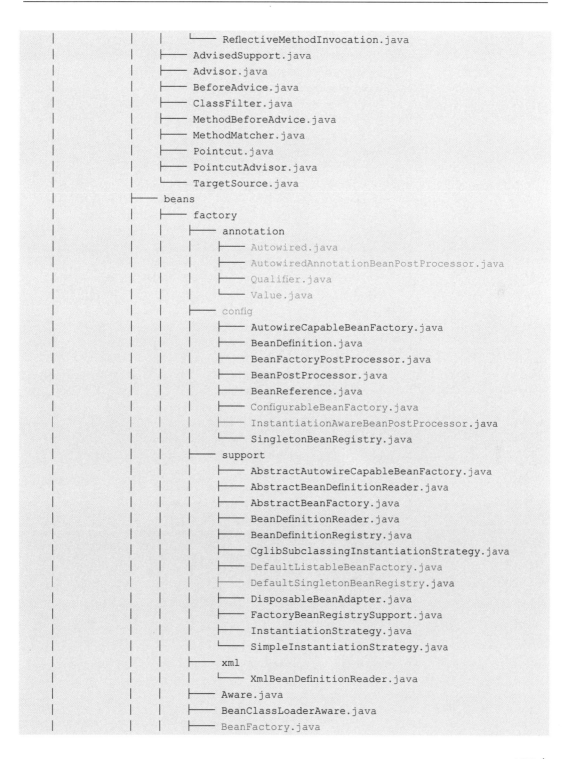

```
│   │   │   ├── BeanFactoryAware.java
│   │   │   ├── BeanNameAware.java
│   │   │   ├── ConfigurableListableBeanFactory.java
│   │   │   ├── DisposableBean.java
│   │   │   ├── FactoryBean.java
│   │   │   ├── HierarchicalBeanFactory.java
│   │   │   ├── InitializingBean.java
│   │   │   ├── ListableBeanFactory.java
│   │   │   └── PropertyPlaceholderConfigurer.java
│   │   ├── BeansException.java
│   │   ├── PropertyValue.java
│   │   └── PropertyValues.java
│   ├── context
│   │   ├── annotation
│   │   │   ├── ClassPathBeanDefinitionScanner.java
│   │   │   ├── ClassPathScanningCandidateComponentProvider.java
│   │   │   └── Scope.java
│   │   ├── event
│   │   │   ├── AbstractApplicationEventMulticaster.java
│   │   │   ├── ApplicationContextEvent.java
│   │   │   ├── ApplicationEventMulticaster.java
│   │   │   ├── ContextClosedEvent.java
│   │   │   ├── ContextRefreshedEvent.java
│   │   │   └── SimpleApplicationEventMulticaster.java
│   │   ├── support
│   │   │   ├── AbstractApplicationContext.java
│   │   │   ├── AbstractRefreshableApplicationContext.java
│   │   │   ├── AbstractXmlApplicationContext.java
│   │   │   ├── ApplicationContextAwareProcessor.java
│   │   │   └── ClassPathXmlApplicationContext.java
│   │   ├── ApplicationContext.java
│   │   ├── ApplicationContextAware.java
│   │   ├── ApplicationEvent.java
│   │   ├── ApplicationEventPublisher.java
│   │   ├── ApplicationListener.java
│   │   └── ConfigurableApplicationContext.java
│   ├── core.io
│   │   ├── ClassPathResource.java
│   │   ├── DefaultResourceLoader.java
│   │   ├── FileSystemResource.java
│   │   ├── Resource.java
│   │   ├── ResourceLoader.java
│   │   └── UrlResource.java
│   ├── stereotype
```

```
|           |           ├── Component.java
|           ├── utils
|           |       ├── ClassUtils.java
|           |       └── StringValueResolver.java
└── test
        └── java
                └── cn.bugstack.springframework.test
                        ├── bean
                        |       ├── IUserService.java
                        |       └── UserService.java
                        └── ApiTest.java
```

自动扫描注入占位符配置和对象的类的关系如图 14-2 所示。

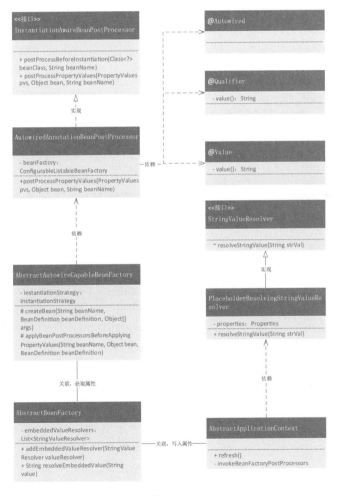

图 14-2

在整个类图中，以围绕实现 InstantiationAwareBeanPostProcessor 接口的类 AutowiredAnnotationBeanPostProcessor 类作为入口点，在使用 AbstractAutowireCapableBeanFactory 类创建 Bean 对象的过程中，调用 BeanPostProcessor 扫描整个类的属性配置中含有自定义注解 @Value、@Autowired、@Qualifier 的属性值。

这里稍微变动的是关于属性值信息的获取方式。将信息注入注解配置的属性字段后，既可以从配置文件中获取占位符，又可以获取 Bean 对象。Bean 对象可以被直接获取。如果想要获取占位符，则需要在 AbstractBeanFactory 中添加新的属性集合 embeddedValueResolvers，并执行 PropertyPlaceholderConfigurer#postProcessBeanFactory 操作将其填充到属性集合中。

2. 将读取到的属性填充到容器

（1）定义解析字符串接口。

源码详见：cn.bugstack.springframework.util.StringValueResolver。

```
public interface StringValueResolver {

    String resolveStringValue(String strVal);

}
```

StringValueResolver 是一个解析字符串操作的接口，由 PlaceholderResolvingStringValueResolver 类实现并完成属性值的获取操作。

（2）填充字符串。

```
public class PropertyPlaceholderConfigurer implements BeanFactoryPostProcessor {

    @Override
    public void postProcessBeanFactory(ConfigurableListableBeanFactory beanFactory) throws BeansException {
        try {
            // 加载属性文件
            DefaultResourceLoader resourceLoader = new DefaultResourceLoader();
            Resource resource = resourceLoader.getResource(location);

            // 替换占位符和设置属性值

            // 向容器中添加字符串解析器，以供解析 @Value 注解使用
            StringValueResolver valueResolver = new PlaceholderResolvingStringValueResolver
```

```
(properties);
            beanFactory.addEmbeddedValueResolver(valueResolver);

    } catch (IOException e) {
        throw new BeansException("Could not load properties", e);
    }
}

private class PlaceholderResolvingStringValueResolver implements StringValueResolver {

    private final Properties properties;

    public PlaceholderResolvingStringValueResolver(Properties properties) {
        this.properties = properties;
    }

    @Override
    public String resolveStringValue(String strVal) {
        return PropertyPlaceholderConfigurer.this.resolvePlaceholder(strVal, properties);
    }

}

}
```

在解析属性配置的 PropertyPlaceholderConfigurer 类中，beanFactory.addEmbeddedValueResolver(valueResolver) 这行代码的作用是将属性值写入 AbstractBeanFactory 类的 embeddedValueResolvers 集合。

这里需要说明的是，embeddedValueResolvers 是 AbstractBeanFactory 类新增加的集合。

3. 自定义注解

自定义注解 @Autowired、@Qualifier、@Value。

```
@Retention(RetentionPolicy.RUNTIME)
@Target({ElementType.CONSTRUCTOR, ElementType.FIELD, ElementType.METHOD})
public @interface Autowired {
}

@Target({ElementType.FIELD, ElementType.METHOD, ElementType.PARAMETER, ElementType.TYPE, ElementType.ANNOTATION_TYPE})
@Retention(RetentionPolicy.RUNTIME)
@Inherited
```

```
@Documented
public @interface Qualifier {

    String value() default "";

}

@Target({ElementType.FIELD, ElementType.METHOD, ElementType.PARAMETER})
@Retention(RetentionPolicy.RUNTIME)
@Documented
public @interface Value {

    /**
     * The actual value expression: e.g. "#{systemProperties.myProp}".
     */
    String value();

}
```

这里定义了 @Autowired、@Qualifier 和 @Value 共 3 个注解。在一般情况下，@Qualifier 注解与 @Autowired 注解配合使用。

4. 扫描自定义注解

源码详见：cn.bugstack.springframework.beans.factory.annotation.AutowiredAnnotationBeanPostProcessor。

```
public class AutowiredAnnotationBeanPostProcessor implements InstantiationAwareBeanPostProcessor, BeanFactoryAware {

    private ConfigurableListableBeanFactory beanFactory;

    @Override
    public void setBeanFactory(BeanFactory beanFactory) throws BeansException {
        this.beanFactory = (ConfigurableListableBeanFactory) beanFactory;
    }

    @Override
    public PropertyValues postProcessPropertyValues(PropertyValues pvs, Object bean, String beanName) throws BeansException {
        // 1. 处理注解 @Value
        Class<?> clazz = bean.getClass();
        clazz = ClassUtils.isCglibProxyClass(clazz) ? clazz.getSuperclass() : clazz;
```

```
        Field[] declaredFields = clazz.getDeclaredFields();

        for (Field field : declaredFields) {
            Value valueAnnotation = field.getAnnotation(Value.class);
            if (null != valueAnnotation) {
                String value = valueAnnotation.value();
                value = beanFactory.resolveEmbeddedValue(value);
                BeanUtil.setFieldValue(bean, field.getName(), value);
            }
        }

        // 2. 处理注解 @Autowired
        for (Field field : declaredFields) {
            Autowired autowiredAnnotation = field.getAnnotation(Autowired.class);
            if (null != autowiredAnnotation) {
                Class<?> fieldType = field.getType();
                String dependentBeanName = null;
                Qualifier qualifierAnnotation = field.getAnnotation(Qualifier.class);
                Object dependentBean = null;
                if (null != qualifierAnnotation) {
                    dependentBeanName = qualifierAnnotation.value();
                    dependentBean = beanFactory.getBean(dependentBeanName, fieldType);
                } else {
                    dependentBean = beanFactory.getBean(fieldType);
                }
                BeanUtil.setFieldValue(bean, field.getName(), dependentBean);
            }
        }

        return pvs;
    }
}
```

AutowiredAnnotationBeanPostProcessor 是在 Bean 对象实例化完成后，一个实现接口 InstantiationAwareBeanPostProcessor 用于设置属性操作前处理属性信息的类和操作的方法。只有实现了 BeanPostProcessor 接口，才能在 Bean 的生命周期中处理初始化信息。

核心方法 postProcessPropertyValues 主要用于处理类中含有 @Value 注解和 @Autowired 注解的属性，并获取和设置属性信息。

这里需要注意的是，因为 AbstractAutowireCapableBeanFactory 是使用 CglibSubclassing InstantiationStrategy 创建的类，所以在 AutowiredAnnotationBeanPostProcessor#postProcess

PropertyValues 中需要使用 Cglib 创建对象，否则不能通过 ClassUtils.isCglibProxyClass(clazz)?clazz.getSuperclass():clazz 这行代码获取类的信息。

5. 在 Bean 的生命周期中调用属性注入

源码详见：cn.bugstack.springframework.beans.factory.support.AbstractAutowireCapableBeanFactory。

```java
public abstract class AbstractAutowireCapableBeanFactory extends AbstractBeanFactory implements AutowireCapableBeanFactory {

    private InstantiationStrategy instantiationStrategy = new CglibSubclassingInstantiationStrategy();

    @Override
    protected Object createBean(String beanName, BeanDefinition beanDefinition, Object[] args) throws BeansException {
        Object bean = null;
        try {
            // 判断是否返回代理 Bean 对象
            bean = resolveBeforeInstantiation(beanName, beanDefinition);
            if (null != bean) {
                return bean;
            }
            // 实例化 Bean 对象
            bean = createBeanInstance(beanDefinition, beanName, args);
            // 在设置 Bean 对象的属性之前，允许 BeanPostProcessor 接口修改属性值
            applyBeanPostProcessorsBeforeApplyingPropertyValues(beanName, bean, beanDefinition);
            // 给 Bean 对象填充属性
            applyPropertyValues(beanName, bean, beanDefinition);
            // 执行 Bean 对象的初始化方法和 BeanPostProcessor 接口的前置和后置处理方法
            bean = initializeBean(beanName, bean, beanDefinition);
        } catch (Exception e) {
            throw new BeansException("Instantiation of bean failed", e);
        }

        // 注册实现了 DisposableBean 接口的 Bean 对象
        registerDisposableBeanIfNecessary(beanName, bean, beanDefinition);

        // 判断 SCOPE_SINGLETON、SCOPE_PROTOTYPE
        if (beanDefinition.isSingleton()) {
```

```
            registerSingleton(beanName, bean);
        }
        return bean;
    }

    /**
     * 在设置 Bean 对象的属性之前，允许 BeanPostProcessor 接口修改属性值
     *
     * @param beanName
     * @param bean
     * @param beanDefinition
     */
    protected void applyBeanPostProcessorsBeforeApplyingPropertyValues(String beanName, Object bean, BeanDefinition beanDefinition) {
        for (BeanPostProcessor beanPostProcessor : getBeanPostProcessors()) {
            if (beanPostProcessor instanceof InstantiationAwareBeanPostProcessor){
                PropertyValues pvs = ((InstantiationAwareBeanPostProcessor) beanPostProcessor).postProcessPropertyValues(beanDefinition.getPropertyValues(), bean, beanName);
                if (null != pvs) {
                    for (PropertyValue propertyValue : pvs.getPropertyValues()) {
                        beanDefinition.getPropertyValues().addPropertyValue(propertyValue);
                    }
                }
            }
        }
    }

    // ...
}
```

在 AbstractAutowireCapableBeanFactory#createBean 方法中新增了一个方法调用，就是在设置 Bean 属性之前，允许 BeanPostProcessor 修改属性值的操作 applyBeanPostProcessorsBeforeApplyingPropertyValues。在 applyBeanPostProcessorsBeforeApplyingPropertyValues 方法中，首先获取已经注入的 BeanPostProcessor 集合，然后从集合中筛选出继承接口 InstantiationAwareBeanPostProcessor 的实现类。最后调用相应的 postProcessPropertyValues 方法及循环设置属性值信息 beanDefinition.getPropertyValues().addPropertyValue(propertyValue)。

14.4 注解使用测试

1. 事先准备

（1）配置 Dao。

```
@Component
public class UserDao {

    private static Map<String, String> hashMap = new HashMap<>();

    static {
        hashMap.put("10001", "小傅哥，北京，亦庄");
        hashMap.put("10002", "八杯水，上海，金山嘴");
        hashMap.put("10003", "阿毛，香港，铜锣湾");
    }

    public String queryUserName(String uId) {
        return hashMap.get(uId);
    }

}
```

给 UserDao 类配置一个自动扫描注册 Bean 对象的注解 @Component，接下来会将这个类注入 UserService 中。

（2）注解注入 UserService。

```
@Component("userService")
public class UserService implements IUserService {

    @Value("${token}")
    private String token;

    @Autowired
    private UserDao userDao;

    public String queryUserInfo() {
        try {
            Thread.sleep(new Random(1).nextInt(100));
```

```
        } catch (InterruptedException e) {
            e.printStackTrace();
        }
        return userDao.queryUserName("10001") + ", " + token;
    }

    // …
}
```

这里包括了两种类型的注入，一种是占位符注入属性信息 @Value("${token}")，另一种是注入对象信息 @Autowired。

2. 属性配置文件

（1）token.properties。

```
token=RejDlI78hu223Opo983Ds
```

（2）spring.xml。

```xml
<?xml version="1.0" encoding="UTF-8"?>
<beans xmlns="http://www.springframework.org/schema/beans"
    xmlns:xsi="http://www.w3.org/2001/XMLSchema-instance"
    xmlns:context="http://www.springframework.org/schema/context"
    xsi:schemaLocation="http://www.springframework.org/schema/beans
        http://www.springframework.org/schema/beans/spring-beans.xsd
      http://www.springframework.org/schema/context">

    <bean class="cn.bugstack.springframework.beans.factory.PropertyPlaceholderConfigurer">
        <property name="location" value="classpath:token.properties"/>
    </bean>

    <context:component-scan base-package="cn.bugstack.springframework.test.bean"/>

</beans>
```

在 spring.xml 配置文件中配置扫描属性信息和自动扫描包的路径范围。

3. 单元测试

```
@Test
public void test_scan() {
    ClassPathXmlApplicationContext applicationContext = new ClassPathXmlApplicationContext("classpath:spring.xml");
    IUserService userService = applicationContext.getBean("userService", IUserService.class);
```

```
        System.out.println(" 测试结果: " + userService.queryUserInfo());
}
```

在进行单元测试时，可以完整地测试将一个类注入 Spring Bean 容器中，同时这个属性信息也可以被自动地扫描填充。

测试结果如图 14-3 所示。

```
测试结果：小傅哥，北京，亦庄，RejDlI78hu223Opo983Ds

Process finished with exit code 0
```

图 14-3

从测试结果中可以看到，通过注解注入属性信息的使用方式已经成功，有自动扫描类和注解注入属性，这与使用 Spring 框架越来越像了。

14.5 本章总结

从注解信息扫描注入的实现内容来看，我们一直在 Bean 的生命周期中进行操作。就像 BeanPostProcessor 用于修改新实例化 Bean 对象的扩展点，提供的接口方法用于处理 Bean 对象实例化前后的操作。因为有时需要做一些差异化的控制，所以还需要继承

BeanPostProcessor 接口，定义新的接口 InstantiationAwareBeanPostProcessor，这样就可以区分不同扩展点的操作。

利用 instanceof 判断接口、利用 field.getAnnotation(Value.class) 获取注解，这些都是在类上标识一些信息，便于使用一些方法找到这些功能，并进行处理。在组件的开发设计中，也可以运用上述方法。我们在思考如何将想要的实现融入一个已经细分好的 Bean 生命周期时，会发现它在任何初始化的时间点上、任何面上，都能进行所需的扩展或改变，体现了程序设计的灵活性。

第 15 章 给代理对象设置属性注入

在提交了开发的需求代码后，似乎该代码整体功能完善、逻辑正确、格式漂亮，但是在测试环节，就会发现各种各样的问题。只有经过反复地调整、修改和完善，才能通过测试。

这是因为在测试环节，测试人员输入的数据并非程序员运行过的简单数据，而是更接近用户真实使用的数据。就像在使用 Spring 框架时，没有人规定用户一定要使用普通的类对象，只要是 Java 的 JDK 提供的各种类操作，就有可能在 Spring 框架下使用。例如，MyBasic 对 Dao 使用了代理类注册到 Spring Bean 容器，当启动 RPC 服务时连接注册中心，分库分表在切面时处理数据源的切换，这些功能流程都需要 Spring 框架来支持才能整合使用。如果在开发过程中没有考虑到这些情况，就可能忽略此类功能的实现，一旦到了测试环节，就会出现问题。

- 本章难度：★★★☆☆
- 本章重点：在处理代理对象的初始化阶段，将这个过程融入 Bean 的生命周期，并通过 TargetSource#getTargetClass 提供的代理对象判断和获取接口，用于反射注入的属性信息。

15.1 代理对象创建过程问题

本章要解决的问题是如何给代理对象中的属性填充相应的值。在将 AOP 动态代理融入 Bean 的生命周期时，代理对象的创建是在创建 Bean 对象之前完成的。也就是说，这个代理对象并没有创建在 Bean 生命周期中。

所以，本章要把代理对象的创建融入 Bean 的生命周期中。也就是说，需要把创建代理对象的逻辑迁移到 Bean 对象执行初始化方法之后，再执行代理对象的创建。

15.2 代理对象属性填充设计

InstantiationAwareBeanPostProcessor 接口原本在 Before 中进行处理，现在需要放到 After 中处理，整体设计如图 15-1 所示。

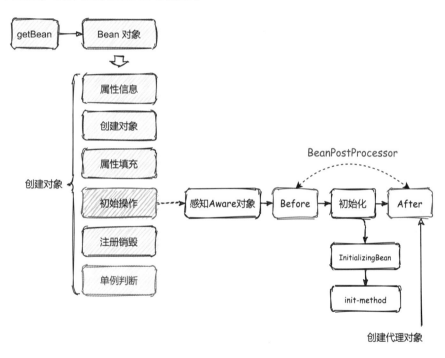

图 15-1

在创建 Bean 对象 createBean 的生命周期中，有一个阶段是在 Bean 对象的属性完成填充以后，执行 Bean 对象的初始化方法，以及 BeanPostProcessor 的前置和后置处理，如感知 Aware 对象、处理 init-method 方法等。这个阶段中的 BeanPostProcessor After 就可以用于创建代理对象。

在 DefaultAdvisorAutoProxyCreator 创建代理对象的过程中，需要把创建操作从 postProcessBeforeInstantiation 方法迁移到 postProcessAfterInitialization 方法，这样才能满足 Bean 属性填充后的创建需求。

15.3　代理对象属性填充实现

1. 工程结构

```
spring-step-15
└── src
    ├── main
    │   └── java
    │       └── cn.bugstack.springframework
    │           ├── aop
    │           │   ├── aspectj
    │           │   │   ├── AspectJExpressionPointcut.java
    │           │   │   └── AspectJExpressionPointcutAdvisor.java
    │           │   ├── framework
    │           │   │   ├── adapter
    │           │   │   │   └── MethodBeforeAdviceInterceptor.java
    │           │   │   ├── autoproxy
    │           │   │   │   └── DefaultAdvisorAutoProxyCreator.java
    │           │   │   ├── AopProxy.java
    │           │   │   ├── Cglib2AopProxy.java
    │           │   │   ├── JdkDynamicAopProxy.java
    │           │   │   ├── ProxyFactory.java
    │           │   │   └── ReflectiveMethodInvocation.java
    │           │   ├── AdvisedSupport.java
    │           │   ├── Advisor.java
    │           │   ├── BeforeAdvice.java
    │           │   ├── ClassFilter.java
    │           │   ├── MethodBeforeAdvice.java
    │           │   ├── MethodMatcher.java
    │           │   ├── Pointcut.java
    │           │   ├── PointcutAdvisor.java
    │           │   └── TargetSource.java
    │           ├── beans
    │           │   ├── factory
    │           │   │   ├── annotation
    │           │   │   │   ├── Autowired.java
```

第15章 给代理对象设置属性注入

```
|   |   |   |   ├── AutowiredAnnotationBeanPostProcessor.java
|   |   |   |   ├── Qualifier.java
|   |   |   |   └── Value.java
|   |   |   ├── config
|   |   |   |   ├── AutowireCapableBeanFactory.java
|   |   |   |   ├── BeanDefinition.java
|   |   |   |   ├── BeanFactoryPostProcessor.java
|   |   |   |   ├── BeanPostProcessor.java
|   |   |   |   ├── BeanReference.java
|   |   |   |   ├── ConfigurableBeanFactory.java
|   |   |   |   ├── InstantiationAwareBeanPostProcessor.java
|   |   |   |   └── SingletonBeanRegistry.java
|   |   |   ├── support
|   |   |   |   ├── AbstractAutowireCapableBeanFactory.java
|   |   |   |   ├── AbstractBeanDefinitionReader.java
|   |   |   |   ├── AbstractBeanFactory.java
|   |   |   |   ├── BeanDefinitionReader.java
|   |   |   |   ├── BeanDefinitionRegistry.java
|   |   |   |   ├── CglibSubclassingInstantiationStrategy.java
|   |   |   |   ├── DefaultListableBeanFactory.java
|   |   |   |   ├── DefaultSingletonBeanRegistry.java
|   |   |   |   ├── DisposableBeanAdapter.java
|   |   |   |   ├── FactoryBeanRegistrySupport.java
|   |   |   |   ├── InstantiationStrategy.java
|   |   |   |   └── SimpleInstantiationStrategy.java
|   |   |   ├── xml
|   |   |   |   └── XmlBeanDefinitionReader.java
|   |   |   ├── Aware.java
|   |   |   ├── BeanClassLoaderAware.java
|   |   |   ├── BeanFactory.java
|   |   |   ├── BeanFactoryAware.java
|   |   |   ├── BeanNameAware.java
|   |   |   ├── ConfigurableListableBeanFactory.java
|   |   |   ├── DisposableBean.java
|   |   |   ├── FactoryBean.java
|   |   |   ├── HierarchicalBeanFactory.java
|   |   |   ├── InitializingBean.java
|   |   |   ├── ListableBeanFactory.java
|   |   |   └── PropertyPlaceholderConfigurer.java
|   |   ├── BeansException.java
|   |   ├── PropertyValue.java
|   |   └── PropertyValues.java
|   ├── context
|   |   ├── annotation
```

```
│   │   │   │   ├── ClassPathBeanDefinitionScanner.java
│   │   │   │   ├── ClassPathScanningCandidateComponentProvider.java
│   │   │   │   └── Scope.java
│   │   │   ├── event
│   │   │   │   ├── AbstractApplicationEventMulticaster.java
│   │   │   │   ├── ApplicationContextEvent.java
│   │   │   │   ├── ApplicationEventMulticaster.java
│   │   │   │   ├── ContextClosedEvent.java
│   │   │   │   ├── ContextRefreshedEvent.java
│   │   │   │   └── SimpleApplicationEventMulticaster.java
│   │   │   ├── support
│   │   │   │   ├── AbstractApplicationContext.java
│   │   │   │   ├── AbstractRefreshableApplicationContext.java
│   │   │   │   ├── AbstractXmlApplicationContext.java
│   │   │   │   ├── ApplicationContextAwareProcessor.java
│   │   │   │   └── ClassPathXmlApplicationContext.java
│   │   │   ├── ApplicationContext.java
│   │   │   ├── ApplicationContextAware.java
│   │   │   ├── ApplicationEvent.java
│   │   │   ├── ApplicationEventPublisher.java
│   │   │   ├── ApplicationListener.java
│   │   │   └── ConfigurableApplicationContext.java
│   │   ├── core.io
│   │   │   ├── ClassPathResource.java
│   │   │   ├── DefaultResourceLoader.java
│   │   │   ├── FileSystemResource.java
│   │   │   ├── Resource.java
│   │   │   ├── ResourceLoader.java
│   │   │   └── UrlResource.java
│   │   ├── stereotype
│   │   │   └── Component.java
│   │   └── utils
│   │       ├── ClassUtils.java
│   │       └── StringValueResolver.java
└── test
    └── java
        └── cn.bugstack.springframework.test
            ├── bean
            │   ├── IUserService.java
            │   └── UserService.java
            └── ApiTest.java
```

在 Bean 的生命周期中创建代理对象的类的关系如图 15-2 所示。

第 15 章 给代理对象设置属性注入

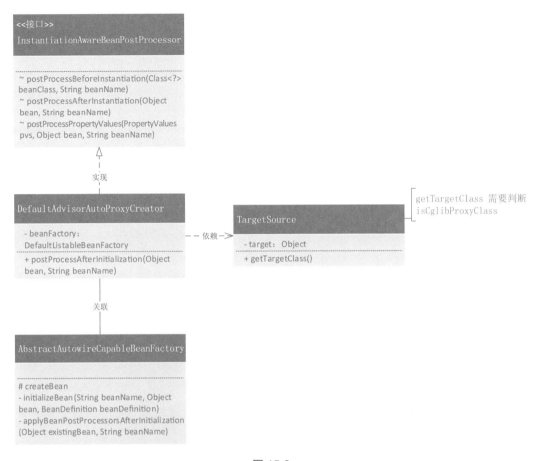

图 15-2

虽然本章要解决的是关于代理对象中属性的填充问题，但实际的解决思路是在 Bean 的生命周期中合适的位置（初始化 initializeBean）处理代理类的创建。

因此，以上的修改并不会涉及太多的内容，主要包括将使用 DefaultAdvisorAutoProxyCreator 类创建代理对象的操作放置在 postProcessAfterInitialization 方法中，以及在对应的 AbstractAutowireCapableBeanFactory 类中完成初始化方法的调用操作。

还有一点需要注意，因为目前在 Spring 框架中，AbstractAutowireCapableBeanFactory 类是使用 CglibSubclassingInstantiationStrategy 来创建对象的，所以当判断对象是否获取接口的方法时，需要判断是否由 Cglib 来创建，否则不能正确获取接口，如 ClassUtils.isCglibProxyClass(clazz)?clazz.getSuperclass():clazz。

2. 判断 Cglib 对象

源码详见：cn.bugstack.springframework.aop.TargetSource。

```
public class TargetSource {

    private final Object target;

    /**
     * Return the type of targets returned by this {@link TargetSource}
     * <p>Can return <code>null</code>, although certain usages of a
     * <code>TargetSource</code> might just work with a predetermined
     * target class
     *
     * @return the type of targets returned by this {@link TargetSource}
     */
    public Class<?>[] getTargetClass() {
        Class<?> clazz = this.target.getClass();
        clazz = ClassUtils.isCglibProxyClass(clazz) ? clazz.getSuperclass() : clazz;
        return clazz.getInterfaces();
    }

}
```

TargetSource#getTargetClass 用于获取 target 对象的接口信息，这个 target 可能是由 JDK 代理创建的，也可能是由 Cglib 创建的。为了正确地获取结果，需要通过 ClassUtils.isCglibProxyClass(clazz) 来判断 Cgtib 对象是否为代理对象，便于找到正确的对象接口。

3. 迁移创建 AOP 代理方法

源码详见：cn.bugstack.springframework.aop.framework.autoproxy.DefaultAdvisorAutoProxyCreator。

```
public class DefaultAdvisorAutoProxyCreator implements InstantiationAwareBeanPostProcessor, BeanFactoryAware {

    private DefaultListableBeanFactory beanFactory;

    @Override
    public Object postProcessBeforeInstantiation(Class<?> beanClass, String beanName) throws BeansException {
        return null;
    }

    @Override
```

```java
public Object postProcessAfterInitialization(Object bean, String beanName) throws BeansException {

    if (isInfrastructureClass(bean.getClass())) return bean;

    Collection<AspectJExpressionPointcutAdvisor> advisors = beanFactory.getBeansOfType(AspectJExpressionPointcutAdvisor.class).values();

    for (AspectJExpressionPointcutAdvisor advisor : advisors) {
        ClassFilter classFilter = advisor.getPointcut().getClassFilter();
        // 过滤匹配类
        if (!classFilter.matches(bean.getClass())) continue;

        AdvisedSupport advisedSupport = new AdvisedSupport();

        TargetSource targetSource = new TargetSource(bean);
        advisedSupport.setTargetSource(targetSource);
        advisedSupport.setMethodInterceptor((MethodInterceptor) advisor.getAdvice());
        advisedSupport.setMethodMatcher(advisor.getPointcut().getMethodMatcher());
        advisedSupport.setProxyTargetClass(false);

        // 返回代理对象
        return new ProxyFactory(advisedSupport).getProxy();

    }

    return bean;
}
```

DefaultAdvisorAutoProxyCreator 类的主要目的是将创建 AOP 代理的操作从 postProcessBeforeInstantiation 方法移动到 postProcessAfterInitialization 方法中。

设置一些必备的 AOP 参数后，返回代理对象 new ProxyFactory(advisedSupport).getProxy。这个代理对象间接调用了 TargetSource 对 getTargetClass 的获取。

4．在 Bean 的生命周期中初始化执行

源码详见：cn.bugstack.springframework.beans.factory.support.AbstractAutowireCapableBeanFactory。

```java
public abstract class AbstractAutowireCapableBeanFactory extends AbstractBeanFactory implements AutowireCapableBeanFactory {
```

```java
    private InstantiationStrategy instantiationStrategy = new CglibSubclassingInstantiationStrategy();

    @Override
    protected Object createBean(String beanName, BeanDefinition beanDefinition, Object[] args) throws BeansException {
        Object bean = null;
        try {
            // ...

            // 执行 Bean 对象的初始化方法和 BeanPostProcessor 接口的前置处理方法和后置处理方法
            bean = initializeBean(beanName, bean, beanDefinition);
        } catch (Exception e) {
            throw new BeansException("Instantiation of bean failed", e);
        }
        // ...
        return bean;
    }

    private Object initializeBean(String beanName, Object bean, BeanDefinition beanDefinition) {

        // ...

        wrappedBean = applyBeanPostProcessorsAfterInitialization(bean, beanName);
        return wrappedBean;
    }

    @Override
    public Object applyBeanPostProcessorsAfterInitialization(Object existingBean, String beanName) throws BeansException {
        Object result = existingBean;
        for (BeanPostProcessor processor : getBeanPostProcessors()) {
            Object current = processor.postProcessAfterInitialization(result, beanName);
            if (null == current) return result;
            result = current;
        }
        return result;
    }

}
```

在 AbstractAutowireCapableBeanFactory#createBean 方法中，重点在于 initializeBean -> applyBeanPostProcessorsAfterInitialization 逻辑的调用，最终完成了 AOP 代理对象的创建操作。

15.4 代理对象属性注入测试

1. 事先准备

UserService 添加属性字段。

```java
public class UserService implements IUserService {

    private String token;

    public String queryUserInfo() {
        try {
            Thread.sleep(new Random(1).nextInt(100));
        } catch (InterruptedException e) {
            e.printStackTrace();
        }
        return "小傅哥, 100001, 深圳, " + token;
    }

}
```

token 是在 UserService 中新增的属性信息, 用于测试代理对象的属性填充操作。

2. 属性配置文件

```xml
<?xml version="1.0" encoding="UTF-8"?>
<beans xmlns="http://www.springframework.org/schema/beans"
       xmlns:xsi="http://www.w3.org/2001/XMLSchema-instance"
       xsi:schemaLocation="http://www.springframework.org/schema/beans
           http://www.springframework.org/schema/beans/spring-beans.xsd">

    <bean id="userService" class="cn.bugstack.springframework.test.bean.UserService">
        <property name="token" value="RejDlI78hu223Opo983Ds"/>
    </bean>

    <bean class="cn.bugstack.springframework.aop.framework.autoproxy.DefaultAdvisorAutoProxyCreator"/>

    <bean id="beforeAdvice" class="cn.bugstack.springframework.test.bean.UserServiceBeforeAdvice"/>

    <bean id="methodInterceptor" class="cn.bugstack.springframework.aop.framework.adapter.MethodBeforeAdviceInterceptor">
```

```xml
        <property name="advice" ref="beforeAdvice"/>
    </bean>

    <bean id="pointcutAdvisor" class="cn.bugstack.springframework.aop.aspectj.AspectJExpressionPointcutAdvisor">
        <property name="expression" value="execution(* cn.bugstack.springframework.test.bean.IUserService.*(..))"/>
        <property name="advice" ref="methodInterceptor"/>
    </bean>

</beans>
```

对于 AOP 的测试来说，唯一新增的就是 property 的配置。

3. 单元测试

```
@Test
public void test_autoProxy() {
    ClassPathXmlApplicationContext applicationContext = new ClassPathXmlApplicationContext("classpath:spring.xml");
    IUserService userService = applicationContext.getBean("userService", IUserService.class);
    System.out.println("测试结果：" + userService.queryUserInfo());
}
```

测试结果如下。

```
拦截方法：queryUserInfo
测试结果：小傅哥, 100001, 深圳, RejDlI78hu223Opo983Ds

Process finished with exit code 0
```

从测试结果中可以看到，我们通过调整 Bean 生命周期，在创建 AOP 代理对象时，已经填充了代理对象的属性信息，如图 15-3 所示。

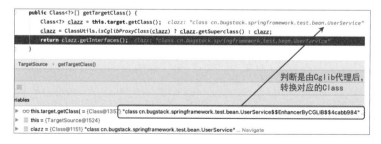

图 15-3

这里还需要注意的是，在 TargetSource#getTargetClass 中需要判断是否由 Cglib 来创建代理对象，只有这样，才可以获取类信息。

15.5 本章总结

本章的核心内容主要是完善 Bean 的生命周期，在创建类的操作中完成代理对象的创建。在创建代理对象的过程中填充相应的属性。

除了实现核心功能，也要关注对象的初始化操作。利用 Cglib 创建对象，会对接口的获取产生影响，因为利用 Cglib 创建对象使用的是 ASM 字节码，所以和普通的 JDK 代理生成的对象不同。

第 16 章

通过三级缓存解决循环依赖

在学习开发 Spring 框架的过程中，相信大家已经体会到我们在开发一些功能时，总是会先思考核心的链路是如何流转的，再不断地扩展新的功能。这与日常的业务代码开发方法相似，通常在一个业务流程中，只有 20% 是核心流程，其余的 80% 流程都是为了保障核心流程顺利执行所做的保护流程和处理机制。

在本书中，处理 AOP 前要先了解代理实现、处理 Aware 前要先了解依赖倒置、处理对象实例化前要先了解 Cglib 等。而本章同样会先通过一个简单的实例来介绍核心内容，再通过处理缓存来解决循环依赖的问题。

- 本章难度：★★★★★
- 本章重点：通过一级缓存学习循环依赖对象的解决方案的核心流程，并在 Spring 框架中添加三级缓存，处理代理对象依赖和填充半成品对象，解决 Bean 对象注入时的循环依赖问题。

16.1 复杂对象的创建思考

目前实现的 Spring 框架是可以满足基本需求的。如果配置了 A、B 两个 Bean 对象互相依赖，那么会出现 java.lang.StackOverflowError 错误提示，为什么呢？因为 A 的创建需要依赖 B 的创建，而 B 的创建又依赖 A 的创建，这样就变成死循环。

这个循环依赖也是 Spring 框架中非常经典的实现，包括以下 3 种情况，如图 16-1 所示。

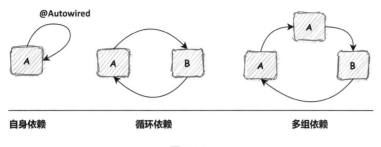

图 16-1

循环依赖主要分为自身依赖、循环依赖和多组依赖。

无论循环依赖的数量有多少，其本质是一样的。即 A 的完整创建依赖于 B，B 的完整创建依赖于 A。但是 A 和 B 之间无法互相解耦，最终导致创建依赖失败。

因此需要 Spring 框架提供除了构造函数注入和原型注入的 setter 循环依赖注入解决方案。

16.2　循环依赖设计

按照 Spring 框架的设计,解决循环依赖需要使用三级缓存,它们分别存储了成品对象、半成品对象（未填充属性值）、代理对象，分阶段存储对象内容，用于解决循环依赖问题。

我们需要知道一个核心的问题：解决循环依赖必须使用三级缓存吗？可以使用二级缓存或者一级缓存吗？其实使用二级缓存或一级缓存都能解决，只是需要注意以下几点：如果只使用一级缓存处理，则流程无法拆分，复杂度也会增加，同时半成品对象可能有空指针异常。如果将半成品对象与成品对象分开，则处理起来更加美观、简单且易扩展。另外，Spring 的两大特性不仅包括 IOC 还包括 AOP，即基于字节码增强后的方法。而三级缓存最主要解决的循环依赖就是对 AOP 的处理。如果把 AOP 代理对象的创建提前，则使用二级缓存也可以解决。但是，这就违背了 Spring 创建对象的原则——Spring 首先将所有的普通 Bean 对象初始化完成，再处理代理对象的初始化。

不过也可以先尝试只使用一级缓存来解决循环依赖问题，通过这种方式可以学习处理循环依赖问题的最核心原理，如图 16-2 所示。

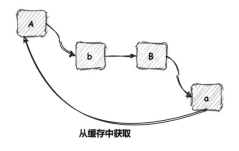

图 16-2

如果只使用一级缓存解决循环依赖，则实现上可以通过在循环依赖 A 对象 newInstance 创建且未填充属性后，直接存储到缓存中。

首先，当使用 A 对象的属性填充 B 对象时，如果在缓存中不能获取 B 对象，则开始创建 B 对象，在创建完成后，把 B 对象填充到缓存中。

然后，对 B 对象的属性进行填充，这时可以从缓存中获取半成品的 A 对象，这时 B 对象的属性就填充完成了。

最后，返回来继续完成 A 对象的属性填充，将其实例化后，填充了属性的 B 对象并赋值给 A 对象的 b 属性，这样就完成了一个循环依赖操作。

代码实现如下。

```
private final static Map<String, Object> singletonObjects = new ConcurrentHashMap<>(256);

private static <T> T getBean(Class<T> beanClass) throws Exception {
    String beanName = beanClass.getSimpleName().toLowerCase();
    if (singletonObjects.containsKey(beanName)) {
        return (T) singletonObjects.get(beanName);
    }
    // 将实例化对象注入缓存
    Object obj = beanClass.newInstance();
    singletonObjects.put(beanName, obj);
    // 填充属性，补全对象
    Field[] fields = obj.getClass().getDeclaredFields();
    for (Field field : fields) {
        field.setAccessible(true);
        Class<?> fieldClass = field.getType();
        String fieldBeanName = fieldClass.getSimpleName().toLowerCase();
        field.set(obj, singletonObjects.containsKey(fieldBeanName) ?
```

```
singletonObjects.get(fieldBeanName) : getBean(fieldClass));
        field.setAccessible(false);
    }
    return (T) obj;
}
```

使用一级缓存存储对象的方式,其实现过程很简单。只要创建完对象,就会立刻注入缓存中。这样就可以在创建其他对象时获取属性需要填充的对象。

测试结果如下。

```
public static void main(String[] args) throws Exception {
    System.out.println(getBean(B.class).getA());
    System.out.println(getBean(A.class).getB());
}

cn.bugstack.springframework.test.A@49476842
cn.bugstack.springframework.test.B@78308db1

Process finished with exit code 0
```

从测试结果和依赖过程中可以看到,使用一级缓存也可以解决简单场景的循环依赖问题,如图 16-3 所示。

图 16-3

getBean 是解决循环依赖问题的核心，当 A 被创建后填充属性时依赖 B，就需要创建 B。在创建完 B 开始填充属性时，发现 B 又依赖 A，因为此时 A 这个半成品对象已经被缓存到 singletonObjects 中，所以 B 可以被正常创建，再通过递归操作创建 A。

先厘清循环依赖的处理过程，再理解循环依赖就没那么复杂了。接下来思考，如果除了简单的对象，还有代理对象和 AOP 应用，那么应如何处理这种依赖问题？整体设计结构如图 16-4 所示。

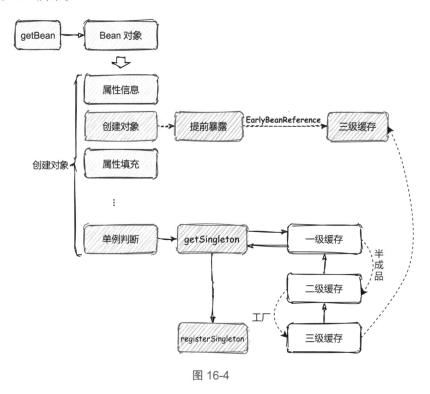

图 16-4

在目前的 Spring 框架中，扩展循环依赖并不会太复杂，主要是创建对象的提前暴露。如果是工厂对象，则使用 getEarlyBeanReference 逻辑提前将对象存储到三级缓存中。后续实际获取的是工厂对象中的 getObject，这才是最终获取的实际对象。

在创建对象的 AbstractAutowireCapableBeanFactory#doCreateBean 方法中，提前暴露对象后，可以通过流程 getSingleton 从 3 个缓存中寻找对象。如果对象在一级缓存、二级缓存中，则可以直接获取该对象；如果对象在三级缓存中，则首先从三级缓存中获取对象，然后删除工厂对象，把实际对象存储到二级缓存中。

关于单例对象的注册操作，就是把实际对象存储到一级缓存中，此时单例对象已经是一个成品对象了。

16.3 循环依赖实现

1. 工程结构

```
|   |   |   |   |   └── Value.java
|   |   |   |   ├── config
|   |   |   |   |   ├── AutowireCapableBeanFactory.java
|   |   |   |   |   ├── BeanDefinition.java
|   |   |   |   |   ├── BeanFactoryPostProcessor.java
|   |   |   |   |   ├── BeanPostProcessor.java
|   |   |   |   |   ├── BeanReference.java
|   |   |   |   |   ├── ConfigurableBeanFactory.java
|   |   |   |   |   ├── InstantiationAwareBeanPostProcessor.java
|   |   |   |   |   └── SingletonBeanRegistry.java
|   |   |   |   ├── support
|   |   |   |   |   ├── AbstractAutowireCapableBeanFactory.java
|   |   |   |   |   ├── AbstractBeanDefinitionReader.java
|   |   |   |   |   ├── AbstractBeanFactory.java
|   |   |   |   |   ├── BeanDefinitionReader.java
|   |   |   |   |   ├── BeanDefinitionRegistry.java
|   |   |   |   |   ├── CglibSubclassingInstantiationStrategy.java
|   |   |   |   |   ├── DefaultListableBeanFactory.java
|   |   |   |   |   ├── DefaultSingletonBeanRegistry.java
|   |   |   |   |   ├── DisposableBeanAdapter.java
|   |   |   |   |   ├── FactoryBeanRegistrySupport.java
|   |   |   |   |   ├── InstantiationStrategy.java
|   |   |   |   |   └── SimpleInstantiationStrategy.java
|   |   |   |   ├── xml
|   |   |   |   |   └── XmlBeanDefinitionReader.java
|   |   |   |   ├── Aware.java
|   |   |   |   ├── BeanClassLoaderAware.java
|   |   |   |   ├── BeanFactory.java
|   |   |   |   ├── BeanFactoryAware.java
|   |   |   |   ├── BeanNameAware.java
|   |   |   |   ├── ConfigurableListableBeanFactory.java
|   |   |   |   ├── DisposableBean.java
|   |   |   |   ├── FactoryBean.java
|   |   |   |   ├── HierarchicalBeanFactory.java
|   |   |   |   ├── InitializingBean.java
|   |   |   |   ├── ListableBeanFactory.java
|   |   |   |   ├── ObjectFactory.java
|   |   |   |   └── PropertyPlaceholderConfigurer.java
|   |   |   ├── BeansException.java
|   |   |   ├── PropertyValue.java
|   |   |   └── PropertyValues.java
|   |   ├── context
|   |   |   ├── annotation
|   |   |   |   ├── ClassPathBeanDefinitionScanner.java
```

第 16 章 通过三级缓存解决循环依赖

```
│   │   │       │   ├── ClassPathScanningCandidateComponentProvider.java
│   │   │       │   └── Scope.java
│   │   │       ├── event
│   │   │       │   ├── AbstractApplicationEventMulticaster.java
│   │   │       │   ├── ApplicationContextEvent.java
│   │   │       │   ├── ApplicationEventMulticaster.java
│   │   │       │   ├── ContextClosedEvent.java
│   │   │       │   ├── ContextRefreshedEvent.java
│   │   │       │   └── SimpleApplicationEventMulticaster.java
│   │   │       ├── support
│   │   │       │   ├── AbstractApplicationContext.java
│   │   │       │   ├── AbstractRefreshableApplicationContext.java
│   │   │       │   ├── AbstractXmlApplicationContext.java
│   │   │       │   ├── ApplicationContextAwareProcessor.java
│   │   │       │   └── ClassPathXmlApplicationContext.java
│   │   │       ├── ApplicationContext.java
│   │   │       ├── ApplicationContextAware.java
│   │   │       ├── ApplicationEvent.java
│   │   │       ├── ApplicationEventPublisher.java
│   │   │       ├── ApplicationListener.java
│   │   │       └── ConfigurableApplicationContext.java
│   │   ├── core.io
│   │   │   ├── ClassPathResource.java
│   │   │   ├── DefaultResourceLoader.java
│   │   │   ├── FileSystemResource.java
│   │   │   ├── Resource.java
│   │   │   ├── ResourceLoader.java
│   │   │   └── UrlResource.java
│   │   ├── stereotype
│   │   │   └── Component.java
│   │   └── utils
│   │       ├── ClassUtils.java
│   │       └── StringValueResolver.java
└── test
    └── java
        └── cn.bugstack.springframework.test
            ├── bean
            │   ├── Husband.java
            │   ├── HusbandMother.java
            │   ├── IMother.java
            │   ├── SpouseAdvice.java
            │   └── Wife.java
            ├── ApiTest.java
            └── CircleTest.java
```

使用三级缓存处理循环依赖核心类的关系如图 16-5 所示。

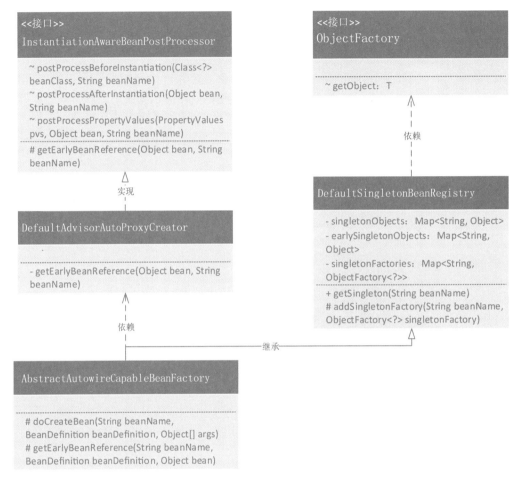

图 16-5

处理循环依赖核心流程类的关系的操作过程如下。

- 循环依赖的核心功能实现主要包括 DefaultSingletonBeanRegistry 提供的三级缓存——singletonObjects、earlySingletonObjects、singletonFactories，分别用于存储成品对象、半成品对象和工厂对象。同时包装 3 个缓存提供方法——getSingleton、registerSingleton、addSingletonFactory，这样用户就可以分别在不同的时间段存储和获取对应的对象。
- AbstractAutowireCapableBeanFactory 中的 doCreateBean 方法提供了关于提前暴露

对象的操作 addSingletonFactory(beanName, () -> getEarlyBeanReference(beanName, beanDefinition, finalBean))，以及后续获取对象和注册对象的操作 exposedObject = getSingleton(beanName)、registerSingleton(beanName, exposedObject)，经过这样处理之后，就可以完成对复杂场景循环依赖的操作。

- 在 DefaultAdvisorAutoProxyCreator 提供的切面服务中，也需要实现 InstantiationAwareBeanPostProcessor 接口中新增的 getEarlyBeanReference 方法，便于把依赖的切面对象也存储到三级缓存中，处理对应的循环依赖。
- 在 Spring 框架中，使用 BeanUtil.setFieldValue(bean, name, value) 来设置对象属性，没有额外添加 Cglib 分支流程的代理判断，避免引入太多的代码。所以，需要将 AbstractAutowireCapableBeanFactory 实例化策略 InstantiationStrategy 修改为 SimpleInstantiationStrategy JDK 方式进行处理。这里可以参考同名称类在 Spring 源码中的处理过程。

2. 设置三级缓存

源码详见：cn.bugstack.springframework.beans.factory.support.DefaultSingletonBeanRegistry。

```
public class DefaultSingletonBeanRegistry implements SingletonBeanRegistry {

    // 一级缓存，普通对象
    private Map<String, Object> singletonObjects = new ConcurrentHashMap<>();

    // 二级缓存，提前暴露对象，没有完全实例化的对象
    protected final Map<String, Object> earlySingletonObjects = new HashMap<String, Object>();

    // 三级缓存，用于存储代理对象
    private final Map<String, ObjectFactory<?>> singletonFactories = new HashMap<String, ObjectFactory<?>>();

    private final Map<String, DisposableBean> disposableBeans = new LinkedHashMap<>();

    @Override
    public Object getSingleton(String beanName) {
        Object singletonObject = singletonObjects.get(beanName);
        if (null == singletonObject) {
            singletonObject = earlySingletonObjects.get(beanName);
            // 判断三级缓存中是否有对象，如果有，则这个对象就是代理对象，因为只有代理对象才会被存储到
            // 三级缓存中
```

```java
            if (null == singletonObject) {
                ObjectFactory<?> singletonFactory = singletonFactories.get(beanName);
                if (singletonFactory != null) {
                    singletonObject = singletonFactory.getObject();
                    // 获取三级缓存中代理对象的真实对象,将其存储到二级缓存中
                    earlySingletonObjects.put(beanName, singletonObject);
                    singletonFactories.remove(beanName);
                }
            }
        }
        return singletonObject;
    }

    public void registerSingleton(String beanName, Object singletonObject) {
        singletonObjects.put(beanName, singletonObject);
        earlySingletonObjects.remove(beanName);
        singletonFactories.remove(beanName);
    }

    protected void addSingletonFactory(String beanName, ObjectFactory<?> singletonFactory){
        if (!this.singletonObjects.containsKey(beanName)) {
            this.singletonFactories.put(beanName, singletonFactory);
            this.earlySingletonObjects.remove(beanName);
        }
    }

    public void registerDisposableBean(String beanName, DisposableBean bean) {
        disposableBeans.put(beanName, bean);
    }
}
```

在用于提供单例对象注册操作的 DefaultSingletonBeanRegistry 类中,共有 3 个缓存对象的属性——singletonObjects、earlySingletonObjects、singletonFactories,分别用于存储不同类型的对象(单例对象、早期的半成品单例对象、单例工厂对象)。

在这 3 个缓存对象中提供了注册、获取和添加不同对象的方法——registerSingleton、getSingleton 和 addSingletonFactory。其中,registerSingleton 方法和 addSingletonFactory 方法都比较简单,getSingleton 方法用于一层一层处理不同时期的单例对象,直至拿到有效的对象。

3. 提前暴露对象

源码详见:cn.bugstack.springframework.beans.factory.support.AbstractAutowireCapable

BeanFactory。

```java
public abstract class AbstractAutowireCapableBeanFactory extends AbstractBeanFactory
implements AutowireCapableBeanFactory {

    protected Object doCreateBean(String beanName, BeanDefinition beanDefinition,
Object[] args) {
        Object bean = null;
        try {
            // 实例化 Bean 对象
            bean = createBeanInstance(beanDefinition, beanName, args);

            // 处理循环依赖，将实例化后的 Bean 对象提前存储到缓存中暴露出来
            if (beanDefinition.isSingleton()) {
                Object finalBean = bean;
                addSingletonFactory(beanName, () -> getEarlyBeanReference(beanName,
beanDefinition, finalBean));
            }

            // 实例化后判断
            boolean continueWithPropertyPopulation = applyBeanPostProcessorsAfterInstantiation
(beanName, bean);
            if (!continueWithPropertyPopulation) {
                return bean;
            }
            // 在设置 Bean 对象的属性之前，允许使用 BeanPostProcessor 接口修改属性值
            applyBeanPostProcessorsBeforeApplyingPropertyValues(beanName, bean,
beanDefinition);
            // 给 Bean 对象填充属性
            applyPropertyValues(beanName, bean, beanDefinition);
            // 执行 Bean 对象的初始化方法和 BeanPostProcessor 接口的前置处理方法和后置处理方法
            bean = initializeBean(beanName, bean, beanDefinition);
        } catch (Exception e) {
            throw new BeansException("Instantiation of bean failed", e);
        }

        // 注册实现了 DisposableBean 接口的 Bean 对象
        registerDisposableBeanIfNecessary(beanName, bean, beanDefinition);

        // 判断 SCOPE_SINGLETON、SCOPE_PROTOTYPE
        Object exposedObject = bean;
        if (beanDefinition.isSingleton()) {
            // 获取代理对象
            exposedObject = getSingleton(beanName);
```

```
            registerSingleton(beanName, exposedObject);
        }
        return exposedObject;
    }

    protected Object getEarlyBeanReference(String beanName, BeanDefinition beanDefinition, Object bean) {
        Object exposedObject = bean;
        for (BeanPostProcessor beanPostProcessor : getBeanPostProcessors()) {
            if (beanPostProcessor instanceof InstantiationAwareBeanPostProcessor) {
                exposedObject = ((InstantiationAwareBeanPostProcessor) beanPostProcessor).getEarlyBeanReference(exposedObject, beanName);
                if (null == exposedObject) return exposedObject;
            }
        }

        return exposedObject;
    }

    // …
}
```

在 AbstractAutowireCapableBeanFactory#doCreateBean 的方法中，主要使用 addSingletonFactory 方法提前暴露了对象、使用 getSingleton 方法获取了单例对象、使用 registerSingleton 方法注册了对象。

getEarlyBeanReference 就是定义在 AOP 切面中的代理对象，可以参考源码中接口 InstantiationAwareBeanPostProcessor#getEarlyBeanReference 的方法的实现过程。

16.4 循环依赖测试

因为要测试循环依赖，所以需要找一个比较真实的场景进行测试。场景中的人物包括：老公和媳妇互相依赖，婆婆代理了媳妇的妈妈的职责，还会通过关怀小两口（切面）来关心家庭生活。

1. 事先准备

老公类。

```java
public class Husband {

    private Wife wife;

    public String queryWife(){
        return "Husband.wife";
    }

}
```

媳妇类。

```java
public class Wife {

    private Husband husband;
    private IMother mother; // 婆婆

    public String queryHusband() {
        return "Wife.husband、Mother.callMother: " + mother.callMother();
    }

}
```

婆婆类，代理了媳妇的妈妈的职责的类。

```java
public class HusbandMother implements FactoryBean<IMother> {

    @Override
    public IMother getObject() throws Exception {
        return (IMother) Proxy.newProxyInstance(Thread.currentThread().getContextClassLoader(), new Class[]{IMother.class}, (proxy, method, args) -> "婚后媳妇妈妈的职责被婆婆代理了！" + method.getName());
    }

}
```

切面类。

```java
public class SpouseAdvice implements MethodBeforeAdvice {

    @Override
    public void before(Method method, Object[] args, Object target) throws Throwable {
        System.out.println("关怀小两口（切面）: " + method);
    }

}
```

2. 属性配置文件

这里的配置就很简单了，首先配置 husband 依赖 wife，然后配置 wife 依赖 husband 和 mother，最后关于 AOP 切面的依赖使用。

```xml
<bean id="husband" class="cn.bugstack.springframework.test.bean.Husband">
    <property name="wife" ref="wife"/>
</bean>

<bean id="wife" class="cn.bugstack.springframework.test.bean.Wife">
    <property name="husband" ref="husband"/>
    <property name="mother" ref="husbandMother"/>
</bean>

<bean id="husbandMother" class="cn.bugstack.springframework.test.bean.HusbandMother"/>

<!-- AOP 配置，验证三级缓存 -->
<bean class="cn.bugstack.springframework.aop.framework.autoproxy.DefaultAdvisorAutoProxyCreator"/>

<bean id="beforeAdvice" class="cn.bugstack.springframework.test.bean.SpouseAdvice"/>

<bean id="methodInterceptor" class="cn.bugstack.springframework.aop.framework.adapter.MethodBeforeAdviceInterceptor">
    <property name="advice" ref="beforeAdvice"/>
</bean>

<bean id="pointcutAdvisor" class="cn.bugstack.springframework.aop.aspectj.AspectJExpressionPointcutAdvisor">
    <property name="expression" value="execution(* cn.bugstack.springframework.test.bean.Wife.*(..))"/>
    <property name="advice" ref="methodInterceptor"/>
</bean>
```

3. 单元测试

```java
@Test
public void test_circular() {
    ClassPathXmlApplicationContext applicationContext = new ClassPathXmlApplicationContext("classpath:spring.xml");
    Husband husband = applicationContext.getBean("husband", Husband.class);
    Wife wife = applicationContext.getBean("wife", Wife.class);
    System.out.println("老公的媳妇：" + husband.queryWife());
    System.out.println("媳妇的老公：" + wife.queryHusband());
}
```

测试结果如图 16-6 所示。

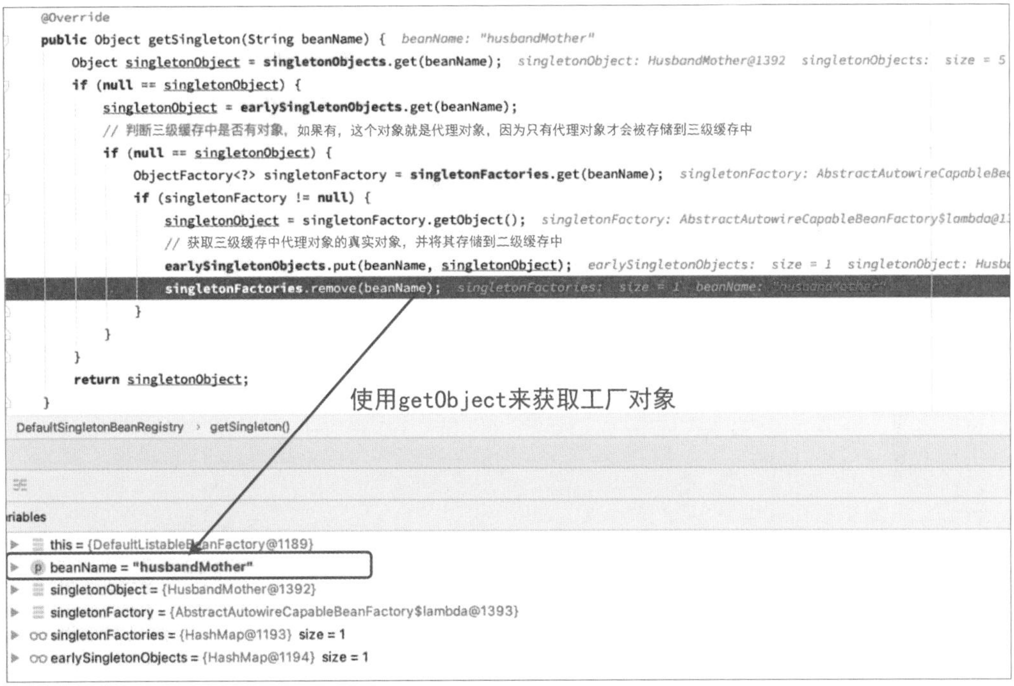

```
@Override
public Object getSingleton(String beanName) {  beanName: "husbandMother"
    Object singletonObject = singletonObjects.get(beanName);  singletonObject: HusbandMother@1392  singletonObjects: size = 5
    if (null == singletonObject) {
        singletonObject = earlySingletonObjects.get(beanName);
        // 判断三级缓存中是否有对象，如果有，这个对象就是代理对象，因为只有代理对象才会被存储到三级缓存中
        if (null == singletonObject) {
            ObjectFactory<?> singletonFactory = singletonFactories.get(beanName);  singletonFactory: AbstractAutowireCapableBe
            if (singletonFactory != null) {
                singletonObject = singletonFactory.getObject();  singletonFactory: AbstractAutowireCapableBeanFactory$lambda@1
                // 获取三级缓存中代理对象的真实对象，并将其存储到二级缓存中
                earlySingletonObjects.put(beanName, singletonObject);  earlySingletonObjects: size = 1  singletonObject: Husba
                singletonFactories.remove(beanName);  singletonFactories: size = 1  beanName: "husbandMother"
            }
        }
    }
    return singletonObject;
}
```

使用getObject来获取工厂对象

- this = {DefaultListableBeanFactory@1189}
- beanName = "husbandMother"
- singletonObject = {HusbandMother@1392}
- singletonFactory = {AbstractAutowireCapableBeanFactory$lambda@1393}
- singletonFactories = {HashMap@1193} size = 1
- earlySingletonObjects = {HashMap@1194} size = 1

图 16-6

测试结果如下。

```
老公的媳妇: Husband.wife
关怀小两口（切面）: public java.lang.String cn.bugstack.springframework.test.bean.Wife.queryHusband()
媳妇的老公: Wife.husband、Mother.callMother: 婚后媳妇妈妈的职责被婆婆代理了！callMother

Process finished with exit code 0
```

从测试结果中可以看到，无论是简单对象依赖：老公依赖媳妇、媳妇依赖老公，还是代理工程对象或者 AOP 切面对象，都可以在三级缓存下解决循环依赖的问题。

此外，从图 16-6 中也可以看到，凡是需要使用三级缓存存储工厂对象的类，都会先使用 getObject 来获取真实对象，并将其存储到半成品对象 earlySingletonObjects 中，再移除工厂对象。

16.5 本章总结

Spring 框架的所有功能都是以解决 Java 编程中的特性而存在的,就像本章处理的循环依赖问题。如果没有 Spring 框架,我们也会尽可能地避免写出循环依赖的操作,因为没有经过加工处理的依赖关系肯定会报错。这也体现了程序从能用到好用的升级。

在解决循环依赖的核心流程中,通过提前暴露对象的设计及创建三级缓存的数据结构来存储不同时期的对象。如果没有切面和工厂中的代理对象,则使用二级缓存也可以获取对象。为了设计上的合理性和可扩展性,创建了三级缓存,用于存储不同时期的对象。

第 17 章 数据类型转换

前面各章已经讲解完 IOC、AOP 在日常使用和学习中频繁出现的技术点。为了补全整个框架的结构，方便读者在学习 Spring 框架时对类型转换的知识有所了解，本章再讲解一些关于此类知识的实现方法。

类型转换也被称为数据转换，如从 String 到 Integer、从 String 到 Date、从 Double 到 Long 等。但是这些操作不需要在已经使用 Spring 框架的情况下进行手动处理，要把这些功能扩展到 Spring 框架中。

- 本章难度：★★★★☆
- 本章重点：提供定义 Java 泛型的类型转换工厂及工厂所需的类型转换服务，并把类型转换服务通过 FactoryBean 注册到 Spring Bean 容器中，为对象中的属性转换提供相应的统一处理方案。

17.1 类型转换设计

如果只是把一个简单的类型转换操作抽象成框架，则仅需要一个标准的接口。谁实现这个接口，谁就具备类型转换的功能。有了这样的接口后，还需要注册、工厂等，才可以把类型转换抽象成一个组件服务。整体设计结构如图 17-1 所示。

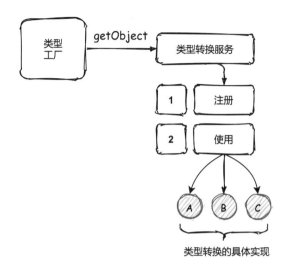

图 17-1

首先从工厂出发，需要实现一个 ConversionServiceFactoryBean 接口，对类型转换服务进行操作。

如果想要实现类型转换，则需要定义 Converter 转换类型、ConverterRegistry 注册类型转换功能。另外，由于类型转换的操作较多，所以需要定义一个类型转换工厂 接口 ConverterFactory，由各个具体的转换操作来实现。

17.2 类型转换实现

1. 工程结构

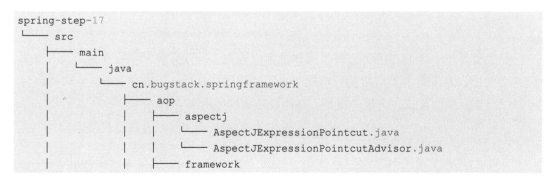

第 17 章 数据类型转换

```
│   │   │       ├── adapter
│   │   │       │   └── MethodBeforeAdviceInterceptor.java
│   │   │       ├── autoproxy
│   │   │       │   └── DefaultAdvisorAutoProxyCreator.java
│   │   │       ├── AopProxy.java
│   │   │       ├── Cglib2AopProxy.java
│   │   │       ├── JdkDynamicAopProxy.java
│   │   │       ├── ProxyFactory.java
│   │   │       └── ReflectiveMethodInvocation.java
│   │   ├── AdvisedSupport.java
│   │   ├── Advisor.java
│   │   ├── BeforeAdvice.java
│   │   ├── ClassFilter.java
│   │   ├── MethodBeforeAdvice.java
│   │   ├── MethodMatcher.java
│   │   ├── Pointcut.java
│   │   ├── PointcutAdvisor.java
│   │   └── TargetSource.java
│   ├── beans
│   │   ├── factory
│   │   │   ├── annotation
│   │   │   │   ├── Autowired.java
│   │   │   │   ├── AutowiredAnnotationBeanPostProcessor.java
│   │   │   │   ├── Qualifier.java
│   │   │   │   └── Value.java
│   │   │   ├── config
│   │   │   │   ├── AutowireCapableBeanFactory.java
│   │   │   │   ├── BeanDefinition.java
│   │   │   │   ├── BeanFactoryPostProcessor.java
│   │   │   │   ├── BeanPostProcessor.java
│   │   │   │   ├── BeanReference.java
│   │   │   │   ├── ConfigurableBeanFactory.java
│   │   │   │   ├── InstantiationAwareBeanPostProcessor.java
│   │   │   │   └── SingletonBeanRegistry.java
│   │   │   ├── support
│   │   │   │   ├── AbstractAutowireCapableBeanFactory.java
│   │   │   │   ├── AbstractBeanDefinitionReader.java
│   │   │   │   ├── AbstractBeanFactory.java
│   │   │   │   ├── BeanDefinitionReader.java
│   │   │   │   ├── BeanDefinitionRegistry.java
│   │   │   │   ├── CglibSubclassingInstantiationStrategy.java
│   │   │   │   ├── DefaultListableBeanFactory.java
│   │   │   │   ├── DefaultSingletonBeanRegistry.java
│   │   │   │   ├── DisposableBeanAdapter.java
```

```
│   │   │   │       ├── FactoryBeanRegistrySupport.java
│   │   │   │       ├── InstantiationStrategy.java
│   │   │   │       └── SimpleInstantiationStrategy.java
│   │   │   ├── xml
│   │   │   │       └── XmlBeanDefinitionReader.java
│   │   │   ├── Aware.java
│   │   │   ├── BeanClassLoaderAware.java
│   │   │   ├── BeanFactory.java
│   │   │   ├── BeanFactoryAware.java
│   │   │   ├── BeanNameAware.java
│   │   │   ├── ConfigurableListableBeanFactory.java
│   │   │   ├── DisposableBean.java
│   │   │   ├── FactoryBean.java
│   │   │   ├── HierarchicalBeanFactory.java
│   │   │   ├── InitializingBean.java
│   │   │   ├── ListableBeanFactory.java
│   │   │   ├── ObjectFactory.java
│   │   │   └── PropertyPlaceholderConfigurer.java
│   │   ├── BeansException.java
│   │   ├── PropertyValue.java
│   │   └── PropertyValues.java
│   ├── context
│   │   ├── annotation
│   │   │   ├── ClassPathBeanDefinitionScanner.java
│   │   │   ├── ClassPathScanningCandidateComponentProvider.java
│   │   │   └── Scope.java
│   │   ├── event
│   │   │   ├── AbstractApplicationEventMulticaster.java
│   │   │   ├── ApplicationContextEvent.java
│   │   │   ├── ApplicationEventMulticaster.java
│   │   │   ├── ContextClosedEvent.java
│   │   │   ├── ContextRefreshedEvent.java
│   │   │   └── SimpleApplicationEventMulticaster.java
│   │   ├── support
│   │   │   ├── AbstractApplicationContext.java
│   │   │   ├── AbstractRefreshableApplicationContext.java
│   │   │   ├── AbstractXmlApplicationContext.java
│   │   │   ├── ApplicationContextAwareProcessor.java
│   │   │   ├── ClassPathXmlApplicationContext.java
│   │   │   └── ConversionServiceFactoryBean.java
│   │   ├── ApplicationContext.java
│   │   ├── ApplicationContextAware.java
│   │   ├── ApplicationEvent.java
│   │   ├── ApplicationEventPublisher.java
```

```
|   |   |       ├── ApplicationListener.java
|   |   |       └── ConfigurableApplicationContext.java
|   |   ├── core
|   |   |   ├── convert
|   |   |   |   ├── converter
|   |   |   |   |   ├── Converter.java
|   |   |   |   |   ├── ConverterFactory.java
|   |   |   |   |   ├── ConverterRegistry.java
|   |   |   |   |   └── GenericConverter.java
|   |   |   |   ├── support
|   |   |   |   |   ├── DefaultConversionService.java
|   |   |   |   |   ├── GenericConversionService.java
|   |   |   |   |   └── StringToNumberConverterFactory.java
|   |   |   |   └── ConversionService.java
|   |   |   └── io
|   |   |       ├── ClassPathResource.java
|   |   |       ├── DefaultResourceLoader.java
|   |   |       ├── FileSystemResource.java
|   |   |       ├── Resource.java
|   |   |       ├── ResourceLoader.java
|   |   |       └── UrlResource.java
|   |   ├── stereotype
|   |   |   └── Component.java
|   |   └── utils
|   |       ├── ClassUtils.java
|   |       ├── NumberUtils.java
|   |       └── StringValueResolver.java
└── test
    └── java
        └── cn.bugstack.springframework.test
            ├── bean
            |   └── Husband.java
            ├── bean
            |   ├── ConvertersFactoryBean.java
            |   ├── StringToIntegerConverter.java
            |   └── StringToLocalDateConverter.java
            └── ApiTest.java
```

数据类型转换处理核心类的关系如图 17-2 所示。

首先，通过添加类型转换接口、类型转换工厂和类型转换的具体操作服务，选择需要被转换的类型，如将字符串类型转换为数值类型。

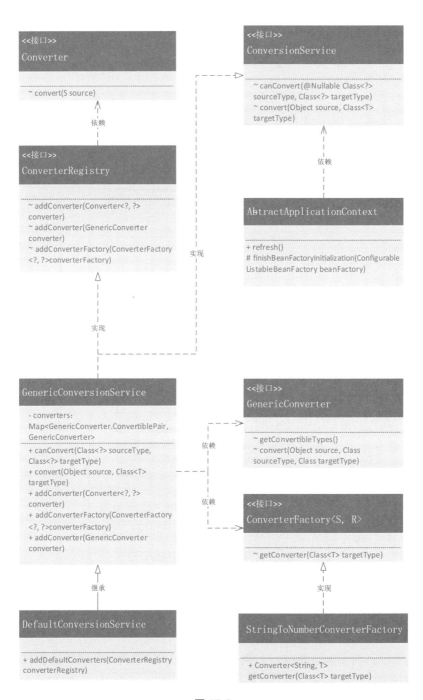

图 17-2

然后，通过与 Spring Bean 工厂的整合把类型转换的服务包装进来，便于配置 Bean 对象的属性信息 applyPropertyValues，在填充属性时可以进行自动转换处理。

2．定义类型转换接口

源码详见（包）：cn.bugstack.springframework.core.convert.Converter。

（1）类型转换处理接口。

```
public interface Converter<S, T> {

    /** Convert the source object of type {@code S} to target type {@code T} */
    T convert(S source);

}
```

（2）类型转换工厂。

```
public interface ConverterFactory<S, R>{

    /**
     * Get the converter to convert from S to target type T, where T is also an instance of R
     * @param <T> the target type
     * @param targetType the target type to convert to
     * @return a converter from S to T
     */
    <T extends R> Converter<S, T> getConverter(Class<T> targetType);

}
```

（3）类型转换注册接口。

```
public interface ConverterRegistry {

    /**
     * Add a plain converter to this registry
     * The convertible source/target type pair is derived from the Converter's parameterized types
     * @throws IllegalArgumentException if the parameterized types could not be resolved
     */
    void addConverter(Converter<?, ?> converter);

    /**
     * Add a generic converter to this registry
     */
```

```
    void addConverter(GenericConverter converter);

    /**
     * Add a ranged converter factory to this registry
     * The convertible source/target type pair is derived from the ConverterFactory's
parameterized types
     * @throws IllegalArgumentException if the parameterized types could not be resolved
     */
    void addConverterFactory(ConverterFactory<?, ?> converterFactory);

}
```

Converter、ConverterFactory、ConverterRegistry 都是用于定义类型转换操作的接口，后续所有的实现都需要围绕这些接口实现，具体的代码功能读者可以进行调试和验证。

3. 实现类型转换服务

源码详见：cn.bugstack.springframework.core.convert.support.DefaultConversionService。

```
public class DefaultConversionService extends GenericConversionService{

    public DefaultConversionService() {
        addDefaultConverters(this);
    }

    public static void addDefaultConverters(ConverterRegistry converterRegistry) {
        // 添加各类的类型转换工厂
        converterRegistry.addConverterFactory(new StringToNumberConverterFactory());
    }

}
```

DefaultConversionService 是继承 GenericConversionService 的实现类，而 GenericConversionService 实现了 ConversionService 和 ConverterRegistry 两个接口，以便 canConvert 进行判断和转换接口 convert 操作。

4. 创建类型转换工厂

源码详见：cn.bugstack.springframework.context.support.ConversionServiceFactoryBean。

```
public class ConversionServiceFactoryBean implements FactoryBean<ConversionService>,
InitializingBean {

    @Nullable
```

```java
    private Set<?> converters;

    @Nullable
    private GenericConversionService conversionService;

    @Override
    public ConversionService getObject() throws Exception {
        return conversionService;
    }

    @Override
    public Class<?> getObjectType() {
        return conversionService.getClass();
    }

    @Override
    public boolean isSingleton() {
        return true;
    }

    @Override
    public void afterPropertiesSet() throws Exception {
        this.conversionService = new DefaultConversionService();
        registerConverters(converters, conversionService);
    }

    private void registerConverters(Set<?> converters, ConverterRegistry registry) {
        if (converters != null) {
            for (Object converter : converters) {
                if (converter instanceof GenericConverter) {
                    registry.addConverter((GenericConverter) converter);
                } else if (converter instanceof Converter<?, ?>) {
                    registry.addConverter((Converter<?, ?>) converter);
                } else if (converter instanceof ConverterFactory<?, ?>) {
                    registry.addConverterFactory((ConverterFactory<?, ?>) converter);
                } else {
                    throw new IllegalArgumentException("Each converter object must implement one of the " + "Converter, ConverterFactory, or GenericConverter interfaces");
                }
            }
        }
    }

    public void setConverters(Set<?> converters) {
```

```
        this.converters = converters;
    }

}
```

有了 FactoryBean 的实现，就可以完成工程对象的操作，还可以提供转换对象的服务 GenericConversionService。另外，在 afterPropertiesSet 中调用了注册转换操作的类，这个类最终会被配置到 spring.xml 配置文件中，在启动过程中会被加载。

5. 使用类型转换服务

源码详见：cn.bugstack.springframework.beans.factory.support.AbstractAutowireCapableBeanFactory。

```
protected void applyPropertyValues(String beanName, Object bean, BeanDefinition beanDefinition) {
    try {
        PropertyValues propertyValues = beanDefinition.getPropertyValues();
        for (PropertyValue propertyValue : propertyValues.getPropertyValues()) {
            String name = propertyValue.getName();
            Object value = propertyValue.getValue();
            if (value instanceof BeanReference) {
                // 例如，A 依赖 B，获取 B 的实例化对象
                BeanReference beanReference = (BeanReference) value;
                value = getBean(beanReference.getBeanName());
            }
            // 类型转换
            else {
                Class<?> sourceType = value.getClass();
                Class<?> targetType = (Class<?>) TypeUtil.getFieldType(bean.getClass(), name);
                ConversionService conversionService = getConversionService();
                if (conversionService != null) {
                    if (conversionService.canConvert(sourceType, targetType)) {
                        value = conversionService.convert(value, targetType);
                    }
                }
            }
            // 反射设置属性填充
            BeanUtil.setFieldValue(bean, name, value);
        }
    } catch (Exception e) {
        throw new BeansException("Error setting property values: " + beanName + "
```

```
        message: " + e);
    }
}
```

在 AbstractAutowireCapableBeanFactory#applyPropertyValues 填充属性的操作中，使用了类型转换的功能。在 AutowiredAnnotationBeanPostProcessor#postProcessPropertyValues 中，也有同样的属性类型转换操作。

17.3 类型转换测试

1. 事先准备

```
public class Husband {

    private String wifiName;

    private Date marriageDate; // 添加一个日期类的转换操作

    // … get/set
}
```

转换时间的操作类。

```
public class StringToLocalDateConverter implements Converter<String, LocalDate> {

    private final DateTimeFormatter DATE_TIME_FORMATTER;

    public StringToLocalDateConverter(String pattern) {
        DATE_TIME_FORMATTER = DateTimeFormatter.ofPattern(pattern);
    }

    @Override
    public LocalDate convert(String source) {
        return LocalDate.parse(source, DATE_TIME_FORMATTER);
    }

}
```

Husband 是一个基础对象类并设置了时间属性，添加一个类型转换的操作，用于转换时间信息。

2. 属性配置文件 spring.xml

首先，配置基础 Bean 对象，设置属性的日期；然后，添加类型转换的服务和自己实现的 ConvertersFactoryBean。

```xml
<bean id="husband" class="cn.bugstack.springframework.test.bean.Husband">
    <property name="wifiName" value="你猜"/>
    <property name="marriageDate" value="2021-08-08"/>
</bean>

<bean id="conversionService" class="cn.bugstack.springframework.context.support.ConversionServiceFactoryBean">
    <property name="converters" ref="converters"/>
</bean>

<bean id="converters" class="cn.bugstack.springframework.test.converter.ConvertersFactoryBean"/>
```

3. 单元测试

```java
@Test
public void test_convert() {
    ClassPathXmlApplicationContext applicationContext = new ClassPathXmlApplicationContext("classpath:spring.xml");
    Husband husband = applicationContext.getBean("husband", Husband.class);
    System.out.println("测试结果: " + husband);
}

@Test
public void test_StringToIntegerConverter() {
    StringToIntegerConverter converter = new StringToIntegerConverter();
    Integer num = converter.convert("1234");
    System.out.println("测试结果: " + num);
}

@Test
public void test_StringToNumberConverterFactory() {
    StringToNumberConverterFactory converterFactory = new StringToNumberConverterFactory();
    Converter<String, Integer> stringToIntegerConverter = converterFactory.getConverter(Integer.class);
    System.out.println("测试结果: " + stringToIntegerConverter.convert("1234"));
    Converter<String, Long> stringToLongConverter = converterFactory.getConverter(Long.class);
    System.out.println("测试结果: " + stringToLongConverter.convert("1234"));
}
```

测试结果如下。

```
测试结果：Husband{wifiName=' 你猜 ', marriageDate=Sun Aug 08 00:00:00 CST 2021}
Process finished with exit code 0
```

这个测试内容比较简单，读者可以自行验证结果。虽然最终的结果看上去比较简单，但是整个框架结构的实现设计还是比较复杂的。将一个转换操作抽象为接口适配、工厂模型等方式是值得借鉴的。

17.4 本章总结

对于类型转换，如果只是简单的功能性的开发需求，则使用 if...else 判断语句就可以完成。但是在一个成熟的框架中，要考虑可复用性、可扩展性，所以需要会使用接口的定义、工厂的使用等方式。

第 18 章 JDBC 功能整合

目前已经完成了对 Bean 对象的生命周期进行管理的核心功能,接下来在这个 Spring 框架下继续扩展 JDBC 的封装应用,方便读者在学习 Spring 源码时,对 Spring 框架的开发和使用有一个更全面的了解。

Spring 的框架结构本身就是一种非常适合扩展的工程底座。我们日常使用的各类框架是非常简单的。这些看上去进行简单配置就能使用的组件,都是在 Spring 提供的容器中扩展的。组件通过把自身的服务对象交给 Spring Bean 容器管理来建立服务关系。

- 本章难度:★★★☆☆
- 本章重点:结合 Spring 框架,封装 JDBC 并对外提供统一的数据库操作模板。读者可以通过这样的方式学习 Spring 框架以外的扩展。

18.1 JdbcTemplate 说明

JDBC 的封装主要实现的是 JdbcTemplate 的功能。Spring 对数据库的操作在 JDBC 上做了深层次的封装。使用 Spring 的注入功能,可以将 DataSource 注册到 JdbcTemplate 中使用。

JdbcTemplate 主要提供了以下几种方法。

- execute 方法:用于执行任何 SQL 语句,一般用于执行 DDL 语句。

- update 方法及 batchUpdate 方法:update 方法用于执行新增、修改、删除等语句,

batchUpdate 方法用于执行与批处理相关的语句。

- query 方法及 queryForXXX 方法：用于执行与查询相关的语句。
- call 方法：用于执行与存储过程、函数相关的语句。

18.2 整合 JDBC 服务设计

Spring 框架将与 JDBC 的相关操作封装在 spring-jdbc 模块下的 JdbcTemplate 类中调用，并结合 Spring 提供的 InitializingBean 接口，在 BeanFactory 设置属性后进行相应的自定义初始化处理，将 JDBC 整合到 Spring 框架中，如图 18-1 所示。

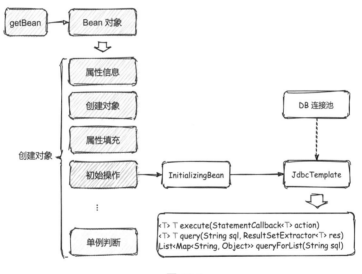

图 18-1

DB 连接池提供数据源服务，这里将 DruidDataSource 作为连接池使用。当然，也可以继续在 Spring 框架中扩展这部分功能，先开发 DriverManagerDataSource 连接池，再进行使用。

在引入连接池后，基于 JdbcTemplate 完成对数据库的操作处理，包括执行 SQL 语句（如查询、更新和删除数据库表等操作，以及开发对应的数据库表语句）。将执行后的结果进行封装，使用 ResultSetExtractor 接口、RowMapper 接口转换数据类型。

18.2 整合 JDBC 服务开发

1. 工程结构

```
spring-step-18
└── src
    ├── main
    │   └── java
    │       └── cn.bugstack.springframework
    │           ├── aop
    │           ├── beans
    │           ├── context
    │           ├── core
    │           ├── jdbc
    │           │   ├── core
    │           │   │   ├── ColumnMapRowMapper.java
    │           │   │   ├── JdbcOperations.java
    │           │   │   ├── JdbcTemplate.java
    │           │   │   ├── ResultSetExtractor.java
    │           │   │   ├── RowMapper.java
    │           │   │   ├── RowMapperResultSetExtractor.java
    │           │   │   ├── SqlProvider.java
    │           │   │   └── StatementCallback.java
    │           │   ├── datasource
    │           │   │   └── DataSourceUtils.java
    │           │   └── support
    │           │       ├── JdbcAccessor.java
    │           │       └── JdbcUtils.java
    │           ├── stereotype
    │           └── utils
    │               ├── ClassUtils.java
    │               ├── NumberUtils.java
    │               └── StringValueResolver.java
    └── test
        └── java
            └── cn.bugstack.springframework.test
                └── ApiTest.java
```

整合 JDBC 功能的核心类的关系如图 18-2 所示。

第 18 章　JDBC 功能整合

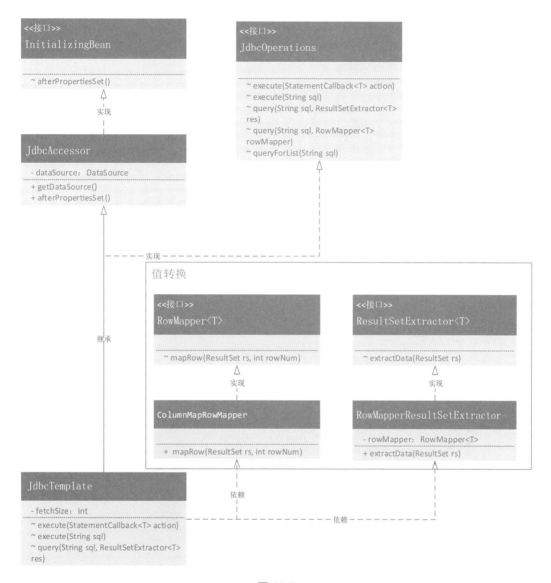

图 18-2

DataSource 用于提供 Connection 链接操作，后续还会扩展辅助类（ConnectionHandle、ConnectionHolder）与数据库事务结合。

JdbcTemplate 是执行数据库操作的入口类，提供 T execute(StatementCallback<T> action, boolean closeResources) 对数据库操作的实现。

JdbcOperations 用于定义很多数据库操作，包括各类的查询处理。这些操作也会调用 execute 方法进行处理，再对数据进行封装。封装操作就是使用 ResultSetExtractor 接口、RowMapper 接口进行数据类型转换的。

2. 数据源操作

源码详见：cn.bugstack.springframework.jdbc.datasource.DataSourceUtils。

```java
public abstract class DataSourceUtils {

    public static Connection getConnection(DataSource dataSource) {
        try {
            return doGetConnection(dataSource);
        } catch (SQLException e) {
            throw new CannotGetJdbcConnectionException("Failed to obtain JDBC Connection", e);
        }
    }

    public static Connection doGetConnection(DataSource dataSource) throws SQLException {
        ConnectionHolder conHolder = (ConnectionHolder) TransactionSynchronizationManager.getResource(dataSource);
        if (null != conHolder && conHolder.hasConnection()) {
            return conHolder.getConnection();
        }
        return fetchConnection(dataSource);
    }

}
```

DataSourceUtils 数据源的操作工具类提供了连接池的链接、关闭、释放等功能，这也是对 Spring 源码的简化。当掌握了所有功能后，可以结合 Spring 源码进行扩展。

3. 数据库执行接口

源码详见：cn.bugstack.springframework.jdbc.core.JdbcOperations。

```java
public interface JdbcOperations {

    <T> T execute(StatementCallback<T> action);

    void execute(String sql);

    <T> T query(String sql, ResultSetExtractor<T> res);
```

```
<T> T query(String sql, Object[] args, ResultSetExtractor<T> rse);

<T> List<T> query(String sql, RowMapper<T> rowMapper);

<T> T query(String sql, PreparedStatementSetter pss, ResultSetExtractor<T> rse);
}
```

在 Spring JDBC 框架中，JdbcOperations 的功能很简单，就是定义了一组用于 JDBC 操作的接口。

4. 数据库操作模板

源码详见：cn.bugstack.springframework.jdbc.core.JdbcTemplate。

```
public class JdbcTemplate extends JdbcAccessor implements JdbcOperations {

    private <T> T execute(StatementCallback<T> action, boolean closeResources) {
        Connection con = DataSourceUtils.getConnection(obtainDataSource());

        Statement stmt = null;
        try {
            stmt = con.createStatement();
            applyStatementSettings(stmt);
            return action.doInStatement(stmt);
        } catch (SQLException e) {
            String sql = getSql(action);
            JdbcUtils.closeStatement(stmt);
            stmt = null;
            throw translateException("ConnectionCallback", sql, e);
        } finally {
            if (closeResources) {
                JdbcUtils.closeStatement(stmt);
            }
        }
    }

    public <T> T query(PreparedStatementCreator psc, final PreparedStatementSetter pss, final ResultSetExtractor<T> rse) {
        Assert.notNull(rse, "ResultSetExtractor must not be null");
        return execute(psc, ps -> {
            ResultSet rs = null;
            try {
                if (pss != null) {
```

```
                pss.setValues(ps);
            }
            rs = ps.executeQuery();
            return rse.extractData(rs);
        } finally {
            JdbcUtils.closeResultSet(rs);
        }
    }, true);
}
```

JdbcTemplate 是对数据库操作的封装，在启动时由外部传入 DataSource 数据源，这也是处理数据库操作最基本的方法。通过这样的封装处理，减少了用户操作的复杂性，也符合设计模式的原则。

execute 方法是整个数据库操作的核心方法，将同类的数据库操作进行统一封装。一些个性化的操作需要进行回调处理。例如，在 JdbcTemplate#query 方法中，也是对 JdbcTemplate#execute 进行包装操作，并返回处理结果。

18.3　JDBC 功能测试

1. 事先准备

SQL 文件：spring-step-18 user.sql。

首先，需要配置一个 MySQL 数据库（8.x 环境）；然后，将存储在源码中的 SQL 语句复制到 MySQL 数据库或 Navicat 等管理工具中。

```
create database spring;

USE spring;

CREATE TABLE `user` (
  `id` bigint(20) NOT NULL AUTO_INCREMENT COMMENT '自增 ID',
  `userId` varchar(9) DEFAULT NULL COMMENT '用户 ID',
  `userHead` varchar(16) DEFAULT NULL COMMENT '用户头像',
  `createTime` datetime DEFAULT NULL COMMENT '创建时间',
  `updateTime` datetime DEFAULT NULL COMMENT '更新时间',
  PRIMARY KEY (`id`)
```

```
) ENGINE=InnoDB AUTO_INCREMENT=2 DEFAULT CHARSET=utf8;
```

还可以直接在 MySQL 控制台或管理界面中执行创建数据库表的操作，这样就可以创建一个用于测试的数据库表了。

2. 配置数据源 spring.xml

在 spring.xml 配置文件中，先配置数据库的链接信息及库表，再将 dataSource 注入 JdbcTemplate 中，由 JdbcTemplate 完成数据库的操作。

```xml
<bean id="dataSource" class="com.alibaba.druid.pool.DruidDataSource">
    <property name="driverClass" value="com.mysql.jdbc.Driver"/>
    <property name="jdbcUrl" value="jdbc:mysql://localhost:3306/spring?useSSL=false"/>
    <property name="username" value="root"/>
    <property name="password" value="123456"/>
</bean>

<bean id="jdbcTemplate"
      class="cn.bugstack.springframework.jdbc.core.JdbcTemplate">
    <property name="dataSource" ref="dataSource"/>
</bean>
```

3. 单元测试

```java
public class ApiTest {

    private JdbcTemplate jdbcTemplate;

    @Before
    public void init() {
        ClassPathXmlApplicationContext applicationContext = new ClassPathXmlApplicationContext("classpath:spring.xml");
        jdbcTemplate = applicationContext.getBean(JdbcTemplate.class);
    }

    @Test
    public void execute(){
        jdbcTemplate.execute("insert into user (id, userId, userHead, createTime, updateTime) values (1, '184172133','01_50', now(), now())");
    }

    @Test
    public void queryForListTest() {
        List<Map<String, Object>> allResult = jdbcTemplate.queryForList("select * from user");
```

```
        for (Map<String, Object> objectMap : allResult) {
            System.out.println("测试结果: " + objectMap);
        }
    }
}
```

测试结果如下。

```
com.alibaba.druid.pool.DruidDataSource info
信息: {dataSource-1} inited
测试结果: {id=1, userId=184172133, userHead=01_50, createTime=2022-03-18 02:11:29.0,
updateTime=2022-03-18 02:11:29.0}

Process finished with exit code 0
```

从测试结果中可以看到，这里已经把操作数据库的 JdbcTemplate 交由 Spring Bean 容器管理，并验证其数据库操作。

18.4 本章总结

本章主要介绍了 Spring Bean 容器的扩展功能，可以在指定的任何阶段把需要交给 Spring 管理的对象进行初始化，如 InitializingBean 可以在 BeanFactory 设置属性后进行相应的处理，整合其他对象。

另外，需要熟悉 JDBC 的包装，使用支撑层 support 承接 Bean 对象的扩展。DataSource 提供了操作数据源的功能，在 core 包中完成对数据库的操作并返回相应的结果。

在学习完本章的内容后，读者可以结合掌握的内容，在 Spring 框架中找到对应的源码进行扩展学习，如 Spring 自身提供的数据源可以实现 DriverManagerDataSource。

第 19 章 事务处理

第 18 章已经对 JDBC 的功能与 Spring 框架进行了整合，使用 Spring 就能操作数据库。关于 Spring 对数据库的操作，还有一个很重要的知识点——事务。

因为 Spring 提供了对数据库事务的管理功能，所以才能让开发者更加方便地通过使用 @Transactional 注解的方式，在切面编程下完成对事务的操作。为了让读者更好地了解 Spring 是如何对事务进行包装的，本章会将重点放在 AOP 切面实现事务控制上。读者通过学习后，就可以更容易地继续学习 Spring 源码了。

- 本章难度：★★★★☆
- 本章重点：运用 @Transactional 注解标识、AOP 切面包装、事务的细分处理，将 JDBC 中的事务操作功能整合到 Spring 的框架结构中。

19.1 了解事务

在了解 Spring 的事务之前，先来介绍一个生活中的实例——在 ATM 上取款。

假设丈夫的银行卡里只有 500 元，出门在外需要现金，需要到 ATM 上取款 500 元。恰好这时妻子使用手机支付买衣服，也需要 500 元，都要从丈夫的银行卡里扣款。在他们同时操作时，妻子支付了 500 元，丈夫是否也能取出 500 元现金呢？显然这是不可能的，因为在事务的控制下，数据必须保证完整性和一致性。这也是事务的 4 个特性——ACID。

- 原子性（Atomicity）：事务是一个原子操作，由一系列的动作组成。事务的原子性确保动作要么全部完成，要么完全不起作用。

- 一致性（Consistency）：一旦事务完成（无论成功还是失败），系统必须确保它所建模的业务处于一致的状态，而不是部分完成、部分失败。现实中的数据不应该被破坏。

- 隔离性（Isolation）：许多事务可能会同时处理相同的数据，因此每个事务都应该与其他事务隔离，防止数据被损坏。

- 持久性（Durability）：一旦事务完成，无论发生什么系统错误，它的结果都不应该受到影响，这样就能从任何系统崩溃中恢复过来。在通常情况下，事务的结果会被写到持久化存储器中。

19.2 事务功能设计

在实现事务代码的逻辑中，我们要尽可能地简化问题单元，注重核心逻辑的实现，避免找不到重点。因此，我们的设计也要先看一下事务的核心逻辑处理该如何完成。

简单来说，我们的最终目的就是把数据库的事务操作用自定义注解和 AOP 切面进行管理，并扩充相应的事务特性。下面先介绍一个最简单的数据库操作事务实例。该数据库操作事务实例的整体结构如图 19-1 所示。

```
@Before
public void init() throws SQLException {
    dataSource = new DruidDataSource();
    dataSource.setDriver(new Driver());
    dataSource.setUrl("jdbc:mysql://localhost:3306/spring?useSSL=false");
    dataSource.setPassword("123456");
    dataSource.setUsername("root");
}

@Test
public void test_translation() throws SQLException {
    connection = dataSource.getConnection().getConnection();
    statement = connection.createStatement();
    connection.setAutoCommit(false);
    try {
```

第 19 章 事务处理

```
        statement.execute("insert into user (id, userId, userHead, createTime,
updateTime) values (1, '184172133','01_50', now(), now())");
        statement.execute("insert into user (id, userId, userHead, createTime,
updateTime) values (1, '184172133','01_50', now(), now())");
    } catch (Exception e) {
        e.printStackTrace();
        connection.rollback();
    }
    connection.commit();
}
```

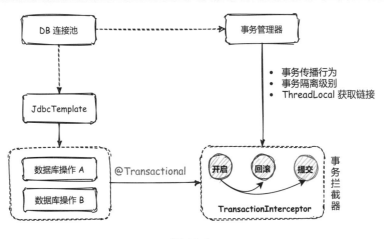

图 19-1

> 注意：关于测试的实例 SQL，已经在本章的工程下提供。

这就是使用 JDBC 直接操作提交事务的方式，首先通过设置自动提交 connection.setAutoCommit(false) 来关闭事务，然后在程序处理过程中进行手动提交事务和回滚事务的操作。

如图 19-1 所示，有了这个结构，就可以把这部分操作数据库事务的功能迁移到 Spring 中，使用注解标记和 AOP 切面进行拦截处理。

如果想要提交事务，则需要保证是在一个链接下进行处理的。这时需要引入事务管理器进行处理，使用 ThreadLocal 进行保存。

只有定义了事务注解，才能对需要处理的事务进行方法拦截。在拦截到事务之后，把这种操作方法交给代理类，也就是把用户对数据库的操作包装到事务的代码库中，方便对操作方法进行事务提交和回滚等操作。

在接下来的实现过程中,不必讲解过多的事务传播行为等操作及其他的辅助类代码,因为太多的代码可能会让读者更难以理解。当逐步了解和掌握核心逻辑后,再扩展其余部分,这时读者就会更容易掌握。

19.3　事务功能实现

1. 工程结构

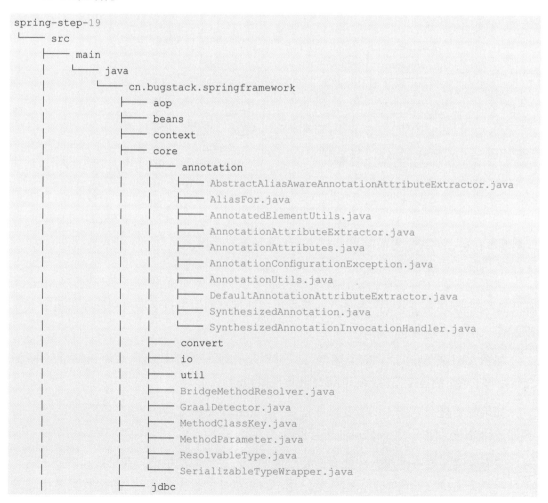

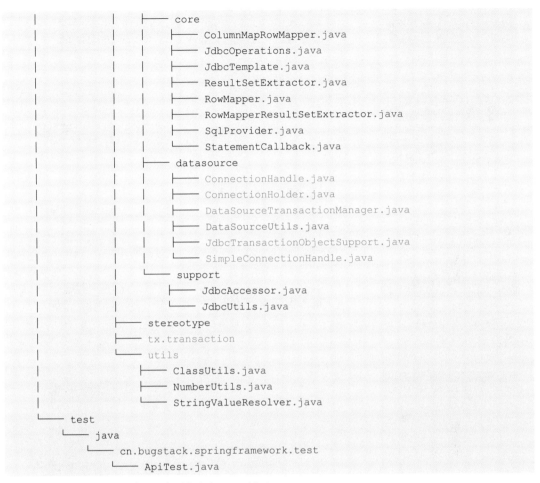

事务管理的核心类的关系如图 19-2 所示。

以自定义事务注解 @Transactional 进行方法标记，用于事务管理扫描。把需要被事务管理的方法拦截 AOP 切面。

因为事务需要在同一个链接下操作，所以需要使用 TransactionSynchronizationManager 类提供的 ThreadLocal 方法记录当前线程下的链接信息。

事务的控制都需要被切面事务支撑类 TransactionAspectSupport 进行包装和处理，以便提交和回滚事务。

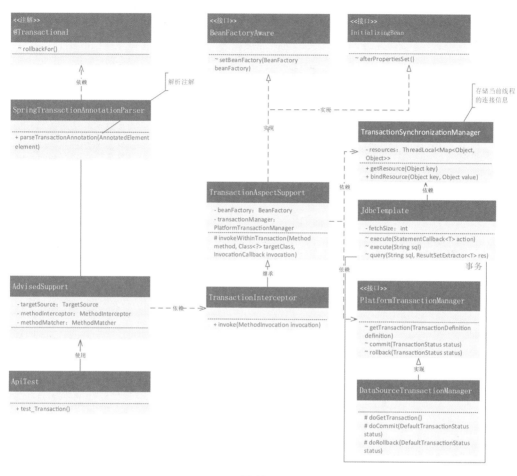

图 19-2

2. 定义事务注解

源码详见：cn.bugstack.springframework.tx.transaction.annotation.Transactional。

```
@Target({ElementType.METHOD, ElementType.TYPE})
@Retention(RetentionPolicy.RUNTIME)
@Inherited
@Documented
public @interface Transactional {

    Class<? extends Throwable>[] rollbackFor() default {};

}
```

只有作用到方法上的事务标记性注解，才能被 SpringTransactionAnnotationParser 类扫描并管理事务。

3. 解析事务注解

源码详见：cn.bugstack.springframework.tx.transaction.annotation.SpringTransactionAnnotationParser。

```
public class SpringTransactionAnnotationParser implements TransactionAnnotationParser, Serializable {

    @Override
    public TransactionAttribute parseTransactionAnnotation(AnnotatedElement element) {
        AnnotationAttributes attributes = AnnotatedElementUtils.findMergedAnnotationAttributes(
                element, Transactional.class, false, false);
        if (null != attributes) {
            return parseTransactionAnnotation(attributes);
        } else {
            return null;
        }
    }

    /**
     * 参照源码，简化实现
     * org.springframework.transaction.annotation.SpringTransactionAnnotationParser#parseTransactionAnnotation
     */
    protected TransactionAttribute parseTransactionAnnotation(AnnotationAttributes attributes) {
        RuleBasedTransactionAttribute rbta = new RuleBasedTransactionAttribute();

        List<RollbackRuleAttribute> rollbackRules = new ArrayList<>();
        for (Class<?> rbRule : attributes.getClassArray("rollbackFor")) {
            rollbackRules.add(new RollbackRuleAttribute(rbRule));
        }

        rbta.setRollbackRules(rollbackRules);
        return rbta;
    }

}
```

SpringTransactionAnnotationParser#parseTransactionAnnotation 方法用于提取配置注解

@Transactional 的方法。这里的提取操作使用了 AnnotatedElementUtils.findMergedAnnotationAttributes 方法，此方法来自 Spring Core 核心包提供的功能。

接下来调用内部的方法 parseTransactionAnnotation(AnnotationAttributes attributes)，用于解析 @Transaction 注解，并设置相关的事务属性。

4．事务拦截处理

（1）方法拦截：MethodInterceptor。

源码详见：cn.bugstack.springframework.tx.transaction.interceptor.TransactionInterceptor。

```
public class TransactionInterceptor extends TransactionAspectSupport implements MethodInterceptor, Serializable {

    @Override
    public Object invoke(MethodInvocation invocation) throws Throwable {
        return invokeWithinTransaction(invocation.getMethod(), invocation.getThis().getClass(), invocation::proceed);
    }

}
```

因为需要使用一个被 AOP 拦截操作处理事务的类，所以需要实现一个事务拦截的功能。

（2）事务处理。

源码详见：cn.bugstack.springframework.tx.transaction.interceptor.TransactionAspectSupport。

```
public abstract class TransactionAspectSupport implements BeanFactoryAware, InitializingBean {

    private static final ThreadLocal<TransactionInfo> transactionInfoHolder =
            new NamedThreadLocal<>("Current aspect-driven transaction");

    private BeanFactory beanFactory;

    private TransactionAttributeSource transactionAttributeSource;

    private PlatformTransactionManager transactionManager;

    protected Object invokeWithinTransaction(Method method, Class<?> targetClass, InvocationCallback invocation) throws Throwable {
        TransactionAttributeSource tas = getTransactionAttributeSource();
        // 查找事务注解 @Transactional
```

```
            TransactionAttribute txAttr = (tas != null ? tas.getTransactionAttribute
(method, targetClass) : null);

        PlatformTransactionManager manager = determineTransactionManager();
        String joinPointIdentification = methodIdentification(method, targetClass);
        TransactionInfo txInfo = createTransactionIfNecessary(manager, txAttr,
joinPointIdentification);

        Object retVal = null;
        try {
            retVal = invocation.proceedWithInvocation();
        } catch (Throwable e) {
            completeTransactionAfterThrowing(txInfo, e);
            throw e;
        } finally {
            cleanupTransactionInfo(txInfo);
        }
        commitTransactionAfterReturning(txInfo);

        return retVal;
    }

}
```

方法拦截器中的 invokeWithinTransaction 方法，就是 TransactionAspectSupport#invokeWithinTransaction 提供的具体拦截操作。

invokeWithinTransaction 方法中的操作包括：提取注解方法、开启事务的提交和回滚，以及把方法包装到 invocation.proceedWithInvocation 中并提交。

到这里，已经完成了最基本的事务注解查找到提交的链路，对于其他的细节，读者可以结合源码进行调试和验证。

5. 事务管理操作

源码详见：cn.bugstack.springframework.tx.transaction.support.AbstractPlatformTransactionManager。

```
public abstract class AbstractPlatformTransactionManager implements PlatformTransactionManager,
Serializable {

    @Override
    public final TransactionStatus getTransaction(TransactionDefinition definition) throws
TransactionException {
```

```java
        Object transaction = doGetTransaction();
        if (null == definition) {
            definition = new DefaultTransactionDefinition();
        }
        if (definition.getTimeout() < TransactionDefinition.TIMEOUT_DEFAULT) {
            throw new TransactionException("Invalid transaction timeout " + definition.getTimeout());
        }
        // 暂定事务传播为默认的行为
        DefaultTransactionStatus status = newTransactionStatus(definition, transaction, true);
        // 开始事务
        doBegin(transaction, definition);
        return status;
    }

    /**
     * 获取事务
     */
    protected abstract Object doGetTransaction() throws TransactionException;

    /**
     * 提交事务
     */
    protected abstract void doCommit(DefaultTransactionStatus status) throws TransactionException;

    /**
     * 回滚事务
     */
    protected abstract void doRollback(DefaultTransactionStatus status) throws TransactionException;

    /**
     * 开始事务
     */
    protected abstract void doBegin(Object transaction, TransactionDefinition definition) throws TransactionException;

}
```

在事务的拦截处理中看到，我们使用 TransactionAspectSupport#invokeWithinTransaction 对事务进行了获取、提交和回滚操作。这些功能都是来自 AbstractPlatformTransactionManager 抽象事务管理器平台类，在这个类中可以设置事务的传播行为、事务的操作。事务的获取、提交、回滚操作也来自数据库连接池提供的功能。

6. 事务同步管理器

（1）使用 ThreadLocal 记录连接信息。

源码详见：cn.bugstack.springframework.tx.transaction.support.TransactionSynchronizationManager。

```java
public abstract class TransactionSynchronizationManager {

    /**
     * 当前线程连接存储
     */
    private static final ThreadLocal<Map<Object, Object>> resources = new NamedThreadLocal<>("Transactional resources");

    /**
     * 事务的名称
     */
    private static final ThreadLocal<String> currentTransactionName = new NamedThreadLocal<>("Current transaction name");

    private static Object doGetResource(Object actualKey) {
        Map<Object, Object> map = resources.get();
        if (null == map) {
            return null;
        }
        return map.get(actualKey);
    }

    public static void bindResource(Object key, Object value) throws IllegalStateException {
        Assert.notNull(value, "Value must not be null");
        Map<Object, Object> map = resources.get();
        if (null == map) {
            map = new HashMap<>();
            resources.set(map);
        }
        map.put(key, value);
    }

}
```

TransactionSynchronizationManager 类主要使用 ThreadLocal 来记录本地线程的特性，保存一次线程下的 DB 链接，便于用户操作数据，以及保证 Spring 管理事务时获取的是同一个链接。否则在不同链接下，就不能再提交事务了。

(2)获取数据库链接。

源码详见:cn.bugstack.springframework.jdbc.datasource.DataSourceUtils。

```java
public abstract class DataSourceUtils {

    public static Connection doGetConnection(DataSource dataSource) throws SQLException {
        ConnectionHolder conHolder = (ConnectionHolder) TransactionSynchronizationManager.getResource(dataSource);
        if (null != conHolder && conHolder.hasConnection()) {
            return conHolder.getConnection();
        }
        return fetchConnection(dataSource);
    }

}
```

在 DataSourceUtils 数据源工具操作类中,最初只是用 dataSource.getConnection 方法获取数据源链接。为了扩充事务的功能,需要使用 TransactionSynchronizationManager.getResource(dataSource) 存储数据源链接,便于操作事务时获取同一个链接。

19.4 切面事务测试

1. 事先准备

SQL 文件:spring-step-18 user.sql。

首先,需要配置一个 MySQL 数据库(8.x 环境);然后,将存储在源码中的 SQL 语句复制到 MySQL 数据库或 Navicat 等管理工具中。

```sql
create database spring;

USE spring;

CREATE TABLE 'user' (
  'id' bigint(20) NOT NULL AUTO_INCREMENT COMMENT '自增 ID',
  'userId' varchar(9) DEFAULT NULL COMMENT '用户 ID',
  'userHead' varchar(16) DEFAULT NULL COMMENT '用户头像',
  'createTime' datetime DEFAULT NULL COMMENT '创建时间',
  'updateTime' datetime DEFAULT NULL COMMENT '更新时间',
  PRIMARY KEY ('id')
```

```
) ENGINE=InnoDB AUTO_INCREMENT=2 DEFAULT CHARSET=utf8;
```

还可以直接在 MySQL 控制台或管理界面中创建数据库表，这样就可以创建数据库表用于测试了。

2. 事务测试类

```java
public class JdbcService {

    /**
     * 使用注解事务
     */
    @Transactional(rollbackFor = Exception.class)
    public void saveData(JdbcTemplate jdbcTemplate) {
        jdbcTemplate.execute("insert into user (id, userId, userHead, createTime, updateTime) values (1, '184172133','01_50', now(), now())");
        jdbcTemplate.execute("insert into user (id, userId, userHead, createTime, updateTime) values (1, '184172133','01_50', now(), now())");
    }

    public void saveDataNoTransaction(JdbcTemplate jdbcTemplate) {
        jdbcTemplate.execute("insert into user (id, userId, userHead, createTime, updateTime) values (1, '184172133','01_50', now(), now())");
        jdbcTemplate.execute("insert into user (id, userId, userHead, createTime, updateTime) values (1, '184172133','01_50', now(), now())");
    }

}
```

提供一个用户操作事务的 Bean 对象有两种方法，一种是使用 @Transactional 注解管理事务，另一种是不使用 @Transactional 注解管理事务。

按照预期，使用事务注解提交两条一样的语句，数据库不会有任何记录，因为存在主键冲突。没有使用事务注解的语句会成功提交一条记录。

3. 配置数据源 spring.xml

```xml
<bean id="dataSource" class="com.alibaba.druid.pool.DruidDataSource">
    <property name="driverClass" value="com.mysql.jdbc.Driver"/>
    <property name="jdbcUrl" value="jdbc:mysql://localhost:3306/spring?useSSL=false"/>
    <property name="username" value="root"/>
    <property name="password" value="123456"/>
</bean>

<bean id="jdbcTemplate"
```

```xml
        class="cn.bugstack.springframework.jdbc.core.JdbcTemplate">
    <property name="dataSource" ref="dataSource"/>
</bean>

<bean id="jdbcService" class="cn.bugstack.springframework.test.bean.JdbcService"/>
```

在 spring.xml 配置文件中，先配置数据库的链接信息及要做的数据库表，再将 dataSource 注入 JdbcTemplate 中，这样使用 JdbcTemplate 就可以完成数据库的操作。

配置 JdbcService Bean 对象是一个用于操作验证数据库事务的方法。

4. 单元测试

```java
public class ApiTest {

    private JdbcTemplate jdbcTemplate;
    private JdbcService jdbcService;
    private DataSource dataSource;

    @Before
    public void init() {
        ClassPathXmlApplicationContext applicationContext = new ClassPathXmlApplicationContext("classpath:spring.xml");
        jdbcTemplate = applicationContext.getBean(JdbcTemplate.class);
        dataSource = applicationContext.getBean(DruidDataSource.class);
        jdbcService = applicationContext.getBean(JdbcService.class);
    }

    @Test
    public void test_Transaction() throws SQLException {
        AnnotationTransactionAttributeSource transactionAttributeSource = new AnnotationTransactionAttributeSource();
        transactionAttributeSource.findTransactionAttribute(jdbcService.getClass());

        DataSourceTransactionManager transactionManager = new DataSourceTransactionManager(dataSource);
        TransactionInterceptor interceptor = new TransactionInterceptor(transactionManager, transactionAttributeSource);

        // 组装代理信息
        AdvisedSupport advisedSupport = new AdvisedSupport();
        advisedSupport.setTargetSource(new TargetSource(jdbcService));
        advisedSupport.setMethodInterceptor(interceptor);
```

```
        advisedSupport.setMethodMatcher(new AspectJExpressionPointcut("execution(* cn.
bugstack.springframework.test.bean.JdbcService.*(..))"));

        // 代理对象（Cglib2AopProxy）
        JdbcService proxy_cglib = (JdbcService) new Cglib2AopProxy(advisedSupport).
getProxy();

        // 测试调用，有事务（不能同时提交两条有主键冲突的数据）
        proxy_cglib.saveData(jdbcTemplate);

        // 测试调用，无事务（当提交两条有主键冲突的数据时，只会成功提交一条数据）
        // proxy_cglib.saveDataNoTransaction(jdbcTemplate);
    }

}
```

init 方法是一个初始化操作，用于获取需要的 Bean 对象。

test_Transaction 方法用于配置切面拦截。

下面分别使用代理类调用 proxy_cglib#saveData、proxy_cglib#saveDataNoTransaction 来验证预期结果。

（1）使用事务注解。

对 proxy_cglib.saveDataNoTransaction(jdbcTemplate) 语句进行注释，测试结果如下。

```
cn.bugstack.springframework.jdbc.UncategorizedSQLException: insert into user (id,
userId, userHead, createTime, updateTime) values (1, '184172133','01_50', now(), now())

    at cn.bugstack.springframework.jdbc.core.JdbcTemplate.execute(JdbcTemplate.java:80)
    at cn.bugstack.springframework.jdbc.core.JdbcTemplate.execute(JdbcTemplate.java:105)
Caused by: java.sql.SQLIntegrityConstraintViolationException: Duplicate entry '1' for
key 'PRIMARY'

Process finished with exit code 255
```

使用事务注解的测试结果为报错主键冲突，数据库中不会有任何记录，如图 19-3 所示。

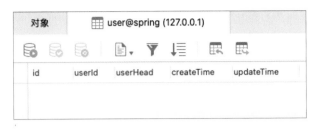

图 19-3

（2）没有使用事务注解。

对 proxy_cglib.saveData(jdbcTemplate) 语句进行注释，测试结果如下。

```
cn.bugstack.springframework.jdbc.UncategorizedSQLException: insert into user (id,
userId, userHead, createTime, updateTime) values (1, '184172133','01_50', now(), now())

    at cn.bugstack.springframework.jdbc.core.JdbcTemplate.execute(JdbcTemplate.java:80)
    at cn.bugstack.springframework.jdbc.core.JdbcTemplate.execute(JdbcTemplate.java:105)
Caused by: java.sql.SQLIntegrityConstraintViolationException: Duplicate entry '1' for
key 'PRIMARY'

Process finished with exit code 255
```

没有使用事务注解同样会出现主键冲突的错误，但是数据库中会添加一条新插入的记录，如图 19-4 所示。

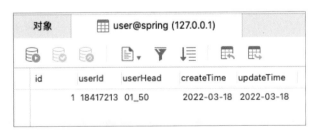

图 19-4

19.5　本章总结

在学习本章内容之前，读者要先了解了事务的特性及其基本的使用流程，再思考如何把事务拆解到 AOP 中进行包装。这个过程也是对一个复杂问题的细化分析和处理，在实际的业务中也有很多这样的问题。

关于事务的操作，我们定义了事务注解，并在 invokeWithinTransaction 方法中使用和提取，并用此方法控制事务，包括获取事务、提交事务、回滚事务等。

ThreadLocal 是一个非常重要的知识点。如果没有使用统一的数据库链接，则无法做到用户的使用链接和切面链接保持为同一个，也就不能提交事务。

相对 Spring 源码，本章对事务的实现简化了很多流程和步骤，读者可以在掌握核心功能后，再学习 Spring 源码。

第 20 章 ORM 框架实现

对象关系映射（Object Relational Mapping，ORM 或 O/RM 或 O/R Mapping）是一种程序设计技术，用于解决面向对象编程语言中不同类型系统的数据之间的转换问题。从使用效果上说，它其实是创建了一个可在编程语言中使用的"虚拟对象数据库"。

从事 Java 开发的程序员几乎都使用过 ORM 框架，如 MyBatis、JPA、Hibernate。将 ORM 框架与 Spring Bean 容器结合使用时，只需要定义 Dao 接口类就可以关联到 XML 或注解配置的数据库操作语句，完成增、删、改、查等功能，这样的功能是如何实现的呢？

本章首先开发一个 ORM 中间件，主要功能是屏蔽底层异构操作，让用户通过一个简单、单一的方式来使用这些服务和功能。这些方式在代码开发中也可以映射在中间件的功能中，这些都是在解决底层的差异，并提供统一的服务。

- 本章难度：★★★☆☆
- 本章重点：读者通过实现 MyBatis 简化版的 ORM 框架，学习文件解析、数据源管理、查询结果封装等核心流程，为下面将 ORM 框架整合到开发的 Spring 框架中做准备。

20.1 简单 ORM 框架设计

如果已经使用过类似 MyBaits、IBatis 的 ORM 框架，那么可以了解到，整个方案设计主要是通过参数映射、SQL 解析和执行及结果封装的方式对数据库进行操作的。

第 20 章 ORM 框架实现

在整个方案的实现过程中，会尽可能地简化实现流程，将 MyBatis 核心的内容展示给读者，便于读者更好地了解一个 ORM 框架的核心部分，再不断地迭代。整体的设计方案如图 20-1 所示。

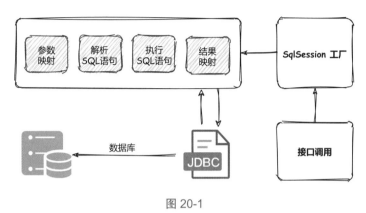

图 20-1

其中，参数映射、解析 SQL 语句、执行 SQL 语句、结果映射都是 ORM 框架的核心内容。这个 ORM 框架会提供调用 SqlSession 工厂的方式。

20.2 简单 ORM 框架实现

1. 工程结构

ORM 中间件的类的关系如图 20-2 所示。

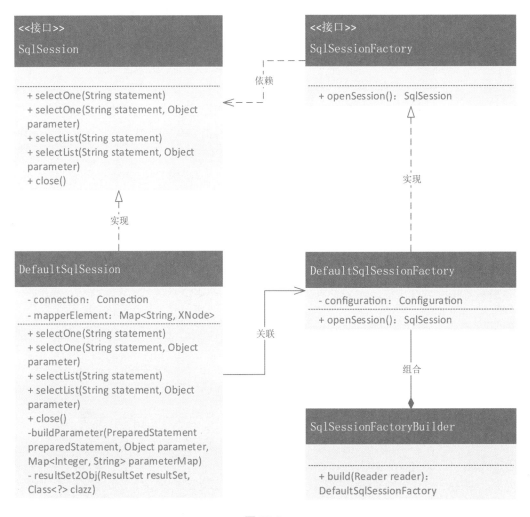

图 20-2

图 20-2 是 ORM 框架实现的核心类的关系，其功能包括加载配置文件，对 XML 进行解析、获取 session 数据库、操作数据库及返回结果。SqlSession 类用于定义和处理数据库，包括常用的方法——T selectOne、List selectList 等。SqlSessionFactory 是开启 session 数据库的工厂处理类，它会操作 DefaultSqlSession。SqlSessionFactoryBuilder 类是数据库操作的核心类，用于处理工厂、解析文件、获取会话信息等。

2．定义 SqlSession 接口

源码详见：cn.bugstack.middleware.mybatis.SqlSession。

```
public interface SqlSession {

    <T> T selectOne(String statement);

    <T> T selectOne(String statement, Object parameter);

    <T> List<T> selectList(String statement);

    <T> List<T> selectList(String statement, Object parameter);

    void close();
}
```

这里定义了数据库操作的查询接口 SqlSession，用于查询一个或多个结果，同时包括有参数和无参数的方法。

3．SqlSession 接口的具体实现类

源码详见：cn.bugstack.middleware.mybatis.DefaultSqlSession.java。

```
public class DefaultSqlSession implements SqlSession {

    private Connection connection;
    private Map<String, XNode> mapperElement;

    public DefaultSqlSession(Connection connection, Map<String, XNode> mapperElement) {
        this.connection = connection;
        this.mapperElement = mapperElement;
    }

    @Override
    public <T> T selectOne(String statement) {
        try {
```

```java
            XNode xNode = mapperElement.get(statement);
            PreparedStatement preparedStatement = connection.prepareStatement(xNode.getSql());
            ResultSet resultSet = preparedStatement.executeQuery();
            List<T> objects = resultSet2Obj(resultSet, Class.forName(xNode.getResultType()));
            return objects.get(0);
        } catch (Exception e) {
            e.printStackTrace();
        }
        return null;
    }

    @Override
    public <T> List<T> selectList(String statement) {
        XNode xNode = mapperElement.get(statement);
        try {
            PreparedStatement preparedStatement = connection.prepareStatement(xNode.getSql());
            ResultSet resultSet = preparedStatement.executeQuery();
            return resultSet2Obj(resultSet, Class.forName(xNode.getResultType()));
        } catch (Exception e) {
            e.printStackTrace();
        }
        return null;
    }

    // ...

    private <T> List<T> resultSet2Obj(ResultSet resultSet, Class<?> clazz) {
        List<T> list = new ArrayList<>();
        try {
            ResultSetMetaData metaData = resultSet.getMetaData();
            int columnCount = metaData.getColumnCount();
            // 每次遍历行值
            while (resultSet.next()) {
                T obj = (T) clazz.newInstance();
                for (int i = 1; i <= columnCount; i++) {
                    Object value = resultSet.getObject(i);
                    String columnName = metaData.getColumnName(i);
                    String setMethod = "set" + columnName.substring(0, 1).toUpperCase() + columnName.substring(1);
                    Method method;
                    if (value instanceof Timestamp) {
                        method = clazz.getMethod(setMethod, Date.class);
                    } else {
                        method = clazz.getMethod(setMethod, value.getClass());
```

```
                }
                method.invoke(obj, value);
            }
            list.add(obj);
        }
    } catch (Exception e) {
        e.printStackTrace();
    }
    return list;
}

@Override
public void close() {
    if (null == connection) return;
    try {
        connection.close();
    } catch (SQLException e) {
        e.printStackTrace();
    }
}
```

这里包括了接口定义的方法实现,即包装了 JDBC 层。通过包装可以隐藏数据库的 JDBC 操作,当调用外部接口时,由内部处理入参和出参。

4. 定义 SqlSessionFactory 接口

源码详见:cn.bugstack.middleware.mybatis.SqlSessionFactory。

```
public interface SqlSessionFactory {

    SqlSession openSession();

}
```

开启一个 SqlSession 接口,这几乎是大家在平时使用时都需要操作的内容。当操作数据库时,会获取每一次执行的 SqlSession。

5. SqlSessionFactory 接口的具体实现类

源码详见:cn.bugstack.middleware.mybatis.DefaultSqlSessionFactory.java。

```
public class DefaultSqlSessionFactory implements SqlSessionFactory {

    private final Configuration configuration;
```

```java
    public DefaultSqlSessionFactory(Configuration configuration) {
        this.configuration = configuration;
    }

    @Override
    public SqlSession openSession() {
        return new DefaultSqlSession(configuration.connection, configuration.mapperElement);
    }

}
```

DefaultSqlSessionFactory 是 MyBatis 最常用的类，这里实现了一个简单版本。虽然是简单的版本，但是也包括了最基本的核心思路。当开启 SqlSession 时，会返回一个 DefaultSqlSession 构造函数。这个构造函数向下传递了 Configuration 配置文件，而配置文件包括 Connection connection、Map dataSource、Map mapperElement。如果用户已经学习了 MyBatis 源码，那么对这一点并不陌生。

6. SqlSessionFactoryBuilder 实现类

源码详见：cn.bugstack.middleware.mybatis.SqlSessionFactoryBuilder.java。

```java
public class SqlSessionFactoryBuilder {

    public DefaultSqlSessionFactory build(Reader reader) {
        SAXReader saxReader = new SAXReader();
        try {
            Document document = saxReader.read(new InputSource(reader));
            Configuration configuration = parseConfiguration(document.getRootElement());
            return new DefaultSqlSessionFactory(configuration);
        } catch (DocumentException e) {
            e.printStackTrace();
        }
        return null;
    }

    public DefaultSqlSessionFactory build(InputStream inputStream) {
        SAXReader saxReader = new SAXReader();
        try {
            Document document = saxReader.read(inputStream);
            Configuration configuration = parseConfiguration(document.getRootElement());
            return new DefaultSqlSessionFactory(configuration);
        } catch (DocumentException e) {
            e.printStackTrace();
```

```java
        }
        return null;
    }

    private Configuration parseConfiguration(Element root) {
        Configuration configuration = new Configuration();
        configuration.setDataSource(dataSource(root.element("environments").element
("environment").element("dataSource")));
        configuration.setConnection(connection(configuration.dataSource));
        configuration.setMapperElement(mapperElement(root.element("mappers")));
        return configuration;
    }

    // 获取数据源配置信息
    private Map<String, String> dataSource(Element element) {
        Map<String, String> dataSource = new HashMap<>(4);
        List<Element> propertyList = element.elements("property");
        for (Element e : propertyList) {
            String name = e.attributeValue("name");
            String value = e.attributeValue("value");
            dataSource.put(name, value);
        }
        return dataSource;
    }

    private Connection connection(Map<String, String> dataSource) {
        try {
            Class.forName(dataSource.get("driver"));
            return DriverManager.getConnection(dataSource.get("url"), dataSource.
get("username"), dataSource.get("password"));
        } catch (ClassNotFoundException | SQLException e) {
            e.printStackTrace();
        }
        return null;
    }

    // 获取 SQL 语句信息
    private Map<String, XNode> mapperElement(Element mappers) {
        Map<String, XNode> map = new HashMap<>();

        List<Element> mapperList = mappers.elements("mapper");
        for (Element e : mapperList) {
            String resource = e.attributeValue("resource");

            try {
                Reader reader = Resources.getResourceAsReader(resource);
```

```java
            SAXReader saxReader = new SAXReader();
            Document document = saxReader.read(new InputSource(reader));
            Element root = document.getRootElement();
            // 命名空间
            String namespace = root.attributeValue("namespace");

            // 执行 SELECT 语句
            List<Element> selectNodes = root.elements("select");
            for (Element node : selectNodes) {
                String id = node.attributeValue("id");
                String parameterType = node.attributeValue("parameterType");
                String resultType = node.attributeValue("resultType");
                String sql = node.getText();

                // 通过正则表达式匹配 SQL 配置中的问号 "?"
                Map<Integer, String> parameter = new HashMap<>();
                Pattern pattern = Pattern.compile("(#\\{(.*?)})");
                Matcher matcher = pattern.matcher(sql);
                for (int i = 1; matcher.find(); i++) {
                    String g1 = matcher.group(1);
                    String g2 = matcher.group(2);
                    parameter.put(i, g2);
                    sql = sql.replace(g1, "?");
                }

                XNode xNode = new XNode();
                xNode.setNamespace(namespace);
                xNode.setId(id);
                xNode.setParameterType(parameterType);
                xNode.setResultType(resultType);
                xNode.setSql(sql);
                xNode.setParameter(parameter);

                map.put(namespace + "." + id, xNode);
            }
        } catch (Exception ex) {
            ex.printStackTrace();
        }

    }
    return map;
}
```

SqlSessionFactoryBuilder 类包括的核心方法有 build（构建实例化元素）、parseConfiguration（解析配置）、dataSource（获取数据库配置）、connection(Map dataSource)（链接数据库）、mapperElement（解析 SQL 语句）。下面分别介绍其中几种核心方法。

（1）build（构建实例化元素）。

build 类主要用于创建解析 XML 文件中的类，以及初始化 SqlSession 工厂中的类 DefaultSqlSessionFactory。需要注意的是，使用 saxReader.setEntityResolver(new XMLMapperEntityResolver()) 代码是为了保证在不联网时同样可以解析 XML 文件中的类，否则需要从互联网中获取 DTD 文件。

（2）parseConfiguration（解析配置）。

获取 XML 文件中的元素，这里主要获取了 dataSource 数据源配置和 mappers 映射语句配置。在这两个配置中，一个是数据库的链接信息，另一个是对数据库操作语句的解析。

（3）connection(Map dataSource)（链接数据库）。

用户可以使用 Class.forName(dataSource.get("driver")) 来链接数据库。这样包装后，外部不需要知道具体的操作。同时，当需要链接多个数据库时，也可以在这里进行扩展设置。

（4）mapperElement（解析 SQL 语句）。

虽然这部分代码块的内容相对较多，但是它的核心是解析 XML 文件中 SQL 语句的配置。在日常使用过程中，都会配置一些 SQL 语句，也有一些入参的占位符。在这里使用正则表达式进行解析操作。

解析完成的 SQL 语句就有了一个名称和 SQL 的映射关系。当操作数据库时，这个组件就可以通过映射关系获取对应的 SQL 语句并进行其他操作。

20.3　ORM 框架使用测试

1. 创建数据库表的信息

SQL 文件：spring-step-18 user.sql。

首先，需要配置一个 MySQL 数据库（8.x 环境）；然后，将存储在源码中的 SQL

语句复制到 MySQL 数据库或 Navicat 等管理工具中。

```
create database spring;

USE spring;

CREATE TABLE 'user' (
  'id' bigint(20) NOT NULL AUTO_INCREMENT COMMENT '自增 ID',
  'userId' varchar(9) DEFAULT NULL COMMENT '用户 ID',
  'userHead' varchar(16) DEFAULT NULL COMMENT '用户头像',
  'createTime' datetime DEFAULT NULL COMMENT '创建时间',
  'updateTime' datetime DEFAULT NULL COMMENT '更新时间',
  PRIMARY KEY ('id')
) ENGINE=InnoDB AUTO_INCREMENT=2 DEFAULT CHARSET=utf8;

BEGIN;
INSERT INTO 'user' VALUES (1, '184172133', '01_50', '2022-03-18 12:25:08', '2022-03-18 12:25:08');
COMMIT;
```

2. 创建 User 用户类和 Dao 接口类

（1）User 用户类。

```
public class User {

    private Long id;
    private String userId;          // 用户 ID
    private String userNickName;    // 昵称
    private String userHead;        // 头像
    private String userPassword;    // 密码
    private Date createTime;        // 创建时间
    private Date updateTime;        // 更新时间

    // … get/set
}
```

（2）Dao 接口类。

```
public interface IUserDao {

    User queryUserInfoById(Long id);

}
```

User 用户类和 Dao 接口类都是基本的数据库信息。读者使用这两个类也可以扩展出自己想测试的方法，或者其他数据库映射类，这与使用MyBatis创建类的方法是相同的。

3. ORM配置文件

（1）数据库连接信息配置。

```xml
<configuration>
    <environments default="development">
        <environment id="development">
            <transactionManager type="JDBC"/>
            <dataSource type="POOLED">
                <property name="driver" value="com.mysql.cj.jdbc.Driver"/>
                <property name="url" value="jdbc:mysql://127.0.0.1:3306/spring?useUnicode=true"/>
                <property name="username" value="root"/>
                <property name="password" value="123456"/>
            </dataSource>
        </environment>
    </environments>

    <mappers>
        <mapper resource="mapper/User_Mapper.xml"/>
    </mappers>
</configuration>
```

这个配置与MyBatis中的配置基本一致，包括数据库的连接池信息及需要引入的Mapper映射文件。

（2）Mapper配置。

```xml
<select id="queryUserInfoById" parameterType="java.lang.Long" resultType="cn.bugstack.middleware.mybatis.test.po.User">
    SELECT id, userId, userNickName, userHead, userPassword, createTime
    FROM user
    where id = #{id}
</select>
```

4. 查询测试

（1）测试实例。

```java
@Test
public void test_queryUserInfoById() {
    String resource = "mybatis-config-datasource.xml";
    Reader reader;
```

```
    try {
        reader = Resources.getResourceAsReader(resource);
        SqlSessionFactory sqlMapper = new SqlSessionFactoryBuilder().build(reader);
        SqlSession session = sqlMapper.openSession();
        try {
            User user = session.selectOne("cn.bugstack.middleware.mybatis.test.dao.
IUserDao.queryUserInfoById", 1L);
            System.out.println(JSON.toJSONString(user));
        } finally {
            session.close();
            reader.close();
        }
    } catch (Exception e) {
        e.printStackTrace();
    }
}
```

（2）测试结果。

```
{"createTime":1577808000000,"id":1,"userHead":"01_50","userId":"184172133",
"userNickName":"小傅哥","userPassword":"123456"}

Process finished with exit code 0
```

从测试结果中可以看到，我们可以通过自己实现的 ORM 框架查询数据库中的数据并将其映射成 Java 对象。另外，这里实现的 ORM 框架还提供了其他的方法，如使用 List selectList(String statement, Object parameter) 方法也可以进行测试验证。

20.4　本章总结

本章实现了 ORM 框架，避免了 JDBC 操作的复杂性。在调用外部接口时，用户可以通过更加简单的方式使用数据库。

ORM 框架的功能是实现核心模块，与 MyBatis 相比其功能实现还很少，但这样可以让读者更好地理解中间件的功能和 ORM 的实现方法。

后续会将 ORM 框架整合到 Spring 框架和 Spring Bean 容器中，这样读者对此类技术就可以有更完整的认识。

第 21 章 将 ORM 框架整合到 Spring Bean 容器中

ORM 框架与 Spring 整合的需求结果就是我们常用的 MyBatis-Spring 框架。

MyBatis-Spring 框架会将 MyBatis 代码无缝地整合到 Spring 中，允许 MyBatis 参与 Spring 的事务管理，创建映射器 Mapper 和 SqlSession，并将其注入 Bean，将 MyBatis 的异常转换为 Spring 的 DataAccessException，实现应用代码不依赖于 MyBatis、Spring 或 MyBatis-Spring。

我们的目标就是将自己实现的 ORM 框架与 Spring 框架结合，并交给 Spring 管理。这里使用最直接和最简单的方式来实现核心代码，让大家更清楚地看到这部分功能的实现逻辑。

- 本章难度：★★★★☆
- 本章重点：通过 MapperScannerConfigurer 扫描配置目录的方式，将包装了 sqlSessionFactory 对数据库操作细节的 Dao 接口类的代理类 MapperFactoryBean 注册到 Spring Bean 容器中，便于使用 MyBatis。

21.1 ORM-Spring 整合设计

如果不需要为使用 MyBatis 的 Dao 接口类写实现类，那么实现这部分接口的数据库操作就需要使用代理类。这部分实现类包括对 sqlSessionFactory 的调用和对 SqlSession

方法的使用。

在 Spring 中正常使用 Dao 时，需要先将 Dao 注入相应的类属性中，再将代理类的 Bean 对象注册到 Spring Bean 容器中，交给 Spring 管理。整体的设计方案如图 21-1 所示。

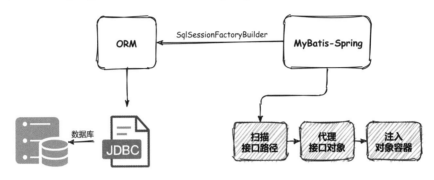

图 21-1

方案设计包括扫描需要注册对象、实现代理类、注册 Bean 对象，这些是将 ORM 与 Spring 结合的核心内容。当所有的内容都被实现后，就可以通过 SqlSessionFactoryBuilder 将 Spring 连接到 ORM 框架。与此同时，还需要在 Spring 框架中添加 BeanDefinitionRegistryPostProcessor 实现，便于将数据库操作 Mapper 的代理对象注入 Spring Bean 容器中。

21.2　ORM-Spring 整合实现

1. 工程结构

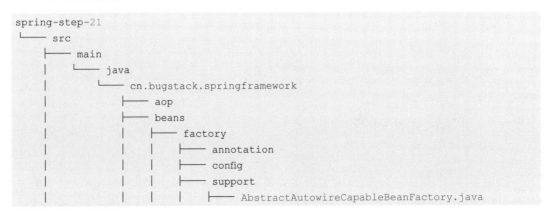

第 21 章　将 ORM 框架整合到 Spring Bean 容器中

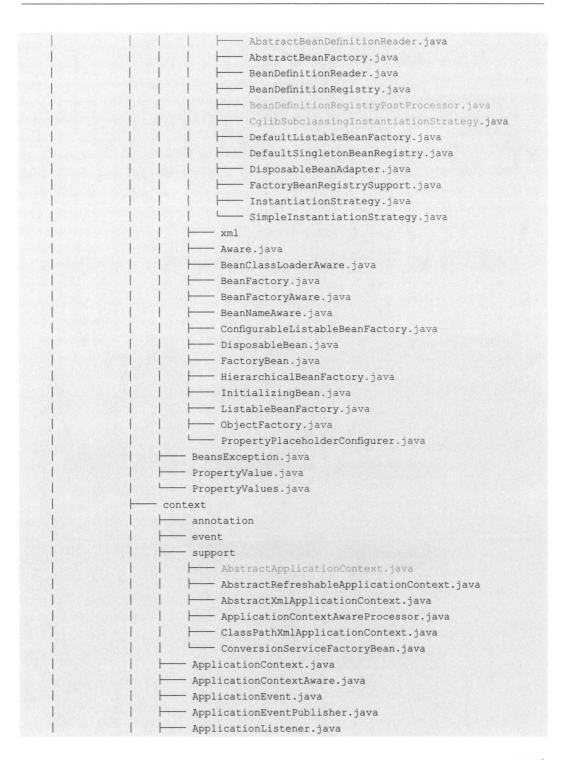

```
  |           |       └── ConfigurableApplicationContext.java
  |           ├── core
  |           ├── jdbc
  |           ├── mybatis
  |           |   ├── MapperFactoryBean.java
  |           |   ├── MapperScannerConfigurer.java
  |           |   └── SqlSessionFactoryBean.java
  |           ├── stereotype
  |           ├── tx.transaction
  |           └── utils
  └── test
      └── java
          └── cn.bugstack.springframework.test
              ├── dao
              |   └── IUserDao.java
              ├── po
              |   └── User.java
              └── ApiTest.java
          └── resources
              ├── mapper
              |   └── User_Mapper.xml
              ├── mybatis-config-datasource.xml
              └── spring-config.xml.xml
```

Spring 整合 ORM 框架的核心类的关系如图 21-2 所示。

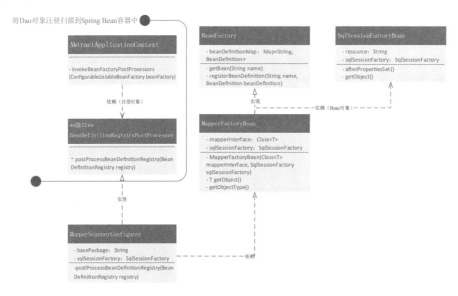

图 21-2

实现 MyBatis-Spring 的核心类主要包括扫描和注入类（MapperScannerConfigurer）、代理类（MapperFactoryBean）。SqlSessionFactoryBean 类是对 SqlSessionFactoryBuilder 类的使用，也是实现 FactoryBean 的一个 Bean 对象。

2．Bean 对象注册接口

先在 Spring 框架中扩展一个允许外部注册对象的接口，这样才能把对 Mapper 创建的代理对象注册到 Spring Bean 容器中。

（1）定义接口。

源码详见：cn.bugstack.springframework.beans.factory.support.BeanDefinitionRegistryPostProcessor。

```
public interface BeanDefinitionRegistryPostProcessor extends BeanFactoryPostProcessor {

    void postProcessBeanDefinitionRegistry(BeanDefinitionRegistry registry) throws BeansException;

}
```

BeanDefinitionRegistryPostProcessor 类继承了 BeanFactoryPostProcessor 类，凡是实现了 BeanDefinitionRegistryPostProcessor 类的接口，即可主动向 Spring Bean 容器注册 Bean 对象。

（2）完成注册。

源码详见：cn.bugstack.springframework.context.support.AbstractApplicationContext。

```
public abstract class AbstractApplicationContext extends DefaultResourceLoader
implements ConfigurableApplicationContext {

    public static final String APPLICATION_EVENT_MULTICASTER_BEAN_NAME =
"applicationEventMulticaster";

    private ApplicationEventMulticaster applicationEventMulticaster;

    @Override
    public void refresh() throws BeansException {
        // 1. 创建 BeanFactory 接口，并加载 BeanDefinition 类
        refreshBeanFactory();

        // 2. 获取 BeanFactory 接口
        ConfigurableListableBeanFactory beanFactory = getBeanFactory();
```

```java
        // 3. 添加 ApplicationContextAwareProcessor 类，让继承自 ApplicationContextAware
        // 的 Bean 对象都能感知所属的 ApplicationContext 接口
        beanFactory.addBeanPostProcessor(new ApplicationContextAwareProcessor(this));

        // 4. 在 Bean 对象实例化之前，执行 BeanFactoryPostProcessor 操作
        invokeBeanFactoryPostProcessors(beanFactory);

    }

    private void invokeBeanFactoryPostProcessors(ConfigurableListableBeanFactory beanFactory) {
        Map<String, BeanFactoryPostProcessor> beanFactoryPostProcessorMap = beanFactory.getBeansOfType(BeanFactoryPostProcessor.class);
        for (BeanFactoryPostProcessor beanFactoryPostProcessor : beanFactoryPostProcessorMap.values()) {
            beanFactoryPostProcessor.postProcessBeanFactory(beanFactory);
        }

        // 注册对象
        if (beanFactory instanceof BeanDefinitionRegistry) {
            BeanDefinitionRegistry registry = (BeanDefinitionRegistry) beanFactory;
            for (BeanFactoryPostProcessor postProcessor : beanFactoryPostProcessorMap.values()) {
                if (postProcessor instanceof BeanDefinitionRegistryPostProcessor) {
                    BeanDefinitionRegistryPostProcessor registryProcessor = (BeanDefinitionRegistryPostProcessor) postProcessor;
                    registryProcessor.postProcessBeanDefinitionRegistry(registry);
                }
            }
        }

}
```

在 AbstractApplicationContext#invokeBeanFactoryPostProcessors 中扩展对 BeanDefinitionRegistryPostProcessor#postProcessBeanDefinitionRegistry 的调用，并注册 Bean 对象，这样就可以将 ORM 中对数据库的操作对象注册进来。

3. SqlSessionFactoryBean 类

源码详见：cn.bugstack.springframework.mybatis.SqlSessionFactoryBean。

```java
public class SqlSessionFactoryBean implements FactoryBean<SqlSessionFactory>,
```

```java
InitializingBean {

    private String resource;
    private SqlSessionFactory sqlSessionFactory;

    @Override
    public void afterPropertiesSet() throws Exception {
        DefaultResourceLoader defaultResourceLoader = new DefaultResourceLoader();
        Resource resource = defaultResourceLoader.getResource(this.resource);

        try (InputStream inputStream = resource.getInputStream()) {
            this.sqlSessionFactory = new SqlSessionFactoryBuilder().build(inputStream);
        } catch (Exception e) {
            e.printStackTrace();
        }
    }

}
```

SqlSessionFactoryBean 类主要实现了 FactoryBean 类、InitializingBean 类，用于加载 MyBatis 核心流程类。实现 InitializingBean 类主要用于加载 MyBatis 相关的内容，如解析 XML、构造 SqlSession、链接数据库等。FactoryBean 类主要包括 getObject 方法、getObjectType 方法和 isSingleton 方法。

4．MapperScannerConfigurer 类

源码详见：cn.bugstack.springframework.mybatis.MapperScannerConfigurer。

```java
public class MapperScannerConfigurer implements BeanDefinitionRegistryPostProcessor {

    private String basePackage;
    private SqlSessionFactory sqlSessionFactory;

    @Override
    public void postProcessBeanDefinitionRegistry(BeanDefinitionRegistry registry) throws BeansException {
        try {
            Set<Class<?>> classes = ClassScanner.scanPackage(basePackage);
            for (Class<?> clazz : classes) {
                // 定义 Bean 对象
                BeanDefinition beanDefinition = new BeanDefinition(clazz);
                PropertyValues propertyValues = new PropertyValues();
                propertyValues.addPropertyValue(new PropertyValue("mapperInterface", clazz));
```

```
                propertyValues.addPropertyValue(new PropertyValue("sqlSessionFactory", 
sqlSessionFactory));
                beanDefinition.setPropertyValues(propertyValues);
                beanDefinition.setBeanClass(MapperFactoryBean.class);
                // 注册 Bean 对象
                registry.registerBeanDefinition(clazz.getSimpleName(), beanDefinition);
            }
        } catch (Exception e) {
            e.printStackTrace();
        }
    }
}
```

MapperScannerConfigurer 类处理的核心内容是将用户配置在 XML 中 Dao 接口类的地址全部扫描出来，并将代理对象注册到 Spring Bean 容器中。首先，通过 ClassScanner.scanPackage(basePackage) 处理类的扫描注册 classpath:org/bugstack/springframework/test/dao/**/.class，解析 class 文件来获取资源信息。然后，在设置 Bean 对象的定义时，也设置 beanDefinition.setBeanClass(MapperFactoryBean.class)，同时在前面设置了相应的属性信息。最后，执行注册操作 registry.registerBeanDefinition(beanName, definitionHolder.getBeanDefinition())。

5．MapperFactoryBean 类

源码详见：cn.bugstack.springframework.mybatis.MapperFactoryBean。

```
public class MapperFactoryBean<T> implements FactoryBean<T> {

    private Class<T> mapperInterface;
    private SqlSessionFactory sqlSessionFactory;

    @Override
    public T getObject() throws Exception {
        InvocationHandler handler = (proxy, method, args) -> {
            // 排除 Object 方法：toString、hashCode
            if (Object.class.equals(method.getDeclaringClass())) {
                return method.invoke(this, args);
            }
            try {
                System.out.println("你被代理了，执行SQL操作！ " + method.getName());
                return sqlSessionFactory.openSession().selectOne(mapperInterface.getName() + "." + method.getName(), args[0]);
            } catch (Exception e) {
```

```
                e.printStackTrace();
            }
            return method.getReturnType().newInstance();
        };
        return (T) Proxy.newProxyInstance(Thread.currentThread().
getContextClassLoader(), new Class[]{mapperInterface}, handler);
    }
}
```

MapperFactoryBean 类非常重要，所有的 Dao 接口类实际上都是它，它对 SQL 的所有操作进行分发处理。为了使代码更加简洁，这里只实现了查询部分，在 MyBatis-Spring 源码中分别执行 select、update、insert、delete 等操作。T getObject 是一个 Java 代理类的实现，这个代理类对象会被注入 Spring Bean 容器中。在调用方法内容时，会执行代理类的实现部分，也就是"你被代理了，执行 SQL 操作！"。InvocationHandler 代理类的实现部分非常简单，主要开启了 SqlSession，并通过固定的 key——cn.bugstack.springframework.test.dao.IUserDao.queryUserInfoById 执行 SQL 操作。返回执行结果后，查询到的结果信息会反射操作成对象类，这里实现的是 ORM 中间件负责的事情。

21.3 整合功能验证

创建数据库表、初始数据、映射对象、XML 配置与 ORM 使用相关的内容已经在第 20 章进行了介绍。如果读者已经掌握了这部分知识，则可以略过。

1. 创建数据库表的信息

SQL 文件：spring-step-18 user.sql。

首先，需要配置一个 MySQL 数据库（8.x 环境）；然后，将存储在源码中的 SQL 语句复制到 MySQL 数据库或 Navicat 等管理工具中。

```sql
create database spring;

USE spring;

CREATE TABLE 'user' (
  'id' bigint(20) NOT NULL AUTO_INCREMENT COMMENT '自增 ID',
  'userId' varchar(9) DEFAULT NULL COMMENT '用户 ID',
  'userHead' varchar(16) DEFAULT NULL COMMENT '用户头像',
```

```
    'createTime' datetime DEFAULT NULL COMMENT '创建时间',
    'updateTime' datetime DEFAULT NULL COMMENT '更新时间',
    PRIMARY KEY ('id')
) ENGINE=InnoDB AUTO_INCREMENT=2 DEFAULT CHARSET=utf8;

BEGIN;
INSERT INTO 'user' VALUES (1, '184172133', '01_50', '2022-03-18 12:25:08', '2022-03-18 12:25:08');
COMMIT;
```

在测试之前,需要准备好数据库表的信息,数据库为 spring,数据表为 user,并初始化用户的一些信息。

2. 创建 User 用户类和 Dao 接口类

(1) User 用户类。

```
public class User {

    private Long id;
    private String userId;        // 用户 ID
    private String userHead;      // 头像
    private Date createTime;      // 创建时间
    private Date updateTime;      // 更新时间

    // … get/set
}
```

(2) Dao 接口类。

```
public interface IUserDao {

    User queryUserInfoById(Long id);

}
```

User 用户类和 Dao 接口类都是基本的数据库信息。读者使用这两个类也可以扩展出自己想测试的方法,或者其他数据库映射类,这与使用 MyBatis 创建类的方法是相同的。

3. ORM 配置文件

(1) 数据库连接信息配置。

```
<configuration>
    <environments default="development">
        <environment id="development">
```

```xml
            <transactionManager type="JDBC"/>
            <dataSource type="POOLED">
                <property name="driver" value="com.mysql.cj.jdbc.Driver"/>
                <property name="url" value="jdbc:mysql://127.0.0.1:3306/spring?useUnicode=true"/>
                <property name="username" value="root"/>
                <property name="password" value="123456"/>
            </dataSource>
        </environment>
    </environments>

    <mappers>
        <mapper resource="mapper/User_Mapper.xml"/>
    </mappers>

</configuration>
```

这个配置与 MyBatis 中的配置基本一致,包括数据库的连接池信息及需要引入的 Mapper 映射文件。

(2) Mapper 配置。

```xml
<select id="queryUserInfoById" parameterType="java.lang.Long" resultType="cn.bugstack.springframework.test.po.User">
    SELECT id, userId, userHead, createTime
    FROM user
    where id = #{id}
</select>
```

4. Spring Config 配置

```xml
<bean id="sqlSessionFactory" class="cn.bugstack.springframework.mybatis.SqlSessionFactoryBean">
    <property name="resource" value="classpath:mybatis-config-datasource.xml"/>
</bean>
<bean class="cn.bugstack.springframework.mybatis.MapperScannerConfigurer">
    <!-- 注入 sqlSessionFactory -->
    <property name="sqlSessionFactory" ref="sqlSessionFactory"/>
    <!-- 给出需要扫描的 Dao 接口包 -->
    <property name="basePackage" value="cn.bugstack.springframework.test.dao"/>
</bean>
```

这部分的配置与 MyBatis-Spring 中的配置基本一致,把 MyBatis 交给 Spring,以及配置相关的扫描和映射关系。

5. 查询测试

(1) 测试实例。

```
@Test
public void test_ClassPathXmlApplicationContext() {
    BeanFactory beanFactory = new ClassPathXmlApplicationContext("spring-config.xml");
    IUserDao userDao = beanFactory.getBean("IUserDao", IUserDao.class);
    User user = userDao.queryUserInfoById(1L);
    logger.info("测试结果：{}", JSON.toJSONString(user));
}
```

(2) 测试结果。

```
你被代理了，执行SQL操作！queryUserInfoById
测试结果：{"createTime":1647624308000,"id":1,"userHead":"01_50","userId":"184172133"}

Process finished with exit code 0
```

从测试结果中可以看到，将实现的 ORM 框架与 Spring 结合的使用已经达到预期效果。

21.4　本章总结

我们在实现 SqlSessionFactoryBean、MapperScannerConfigurer、SqlSessionFactoryBean 这些核心关键类后，可以将 Spring 与 MyBaits 结合起来使用，解决了没有实现类的接口不能处理数据库的增、删、改、查操作的问题。

学习源码技术，可以让我们把很多技术功能复用到中间件的设计和实现中，来解决实际业务场景中遇到的一些通用性问题。